Analytical Measurement Terminology
Handbook of Terms used in Quality Assurance of Analytical Measurement

Analytical Measurement Terminology

Handbook of Terms used in Quality Assurance of Analytical Measurement

John Green, *BP Amoco*
Pete Houlgate, *LGC*
Jim Miller, *University of Loughborough*
Ernie Newman, *Newman Associates*
Geoffrey Phillips OBE, *Consultant*
Alan Rowley, *Alan Rowley Associates*

Coordinating Author
Elizabeth Prichard

LGC, Teddington, UK

ROYAL SOCIETY OF CHEMISTRY

Setting standards
in analytical science

VALID ANALYTICAL MEASUREMENT

A catalogue record for this book is available from the British Library

ISBN 0-85404-443-4

© LGC (Teddington) Limited, 2001

Published for the LGC
by the Royal Society of Chemistry,
Thomas Graham House, Science Park, Milton Road, Cambridge CB4 0WF, UK
Registered Charity Number 207890

For further information see the RSC web site at www.rsc.org

Typeset by Paston PrePress Ltd, Beccles, Suffolk
Printed by Bookcraft Ltd, UK

Preface

Production of this guide was supported under contract with the Department of Trade and Industry as part of the National Measurement System Valid Analytical Measurement (VAM) programme.

The *Handbook* was prepared by staff at LGC in collaboration with members of the VAM Education and Training Working Group whose assistance is gratefully acknowledged.

Education and training matters have been regarded as an essential aspect of the VAM Programme since its inception in 1988. Clearly the reliability of any measurement depends not only on the availability of instrumentation and methodology but also on the knowledge, skills and depth of understanding of the analyst. It has been clear to members of the Working Group for some time that the plethora of terminology used in analytical measurements is the source of much confusion, particularly as many widely used definitions are themselves confusing, even to the experienced analyst. It is our hope that the straight-forward descriptions and examples provided in this *Handbook* will be of assistance to a wide variety of new and experienced practitioners as well as to teachers and lecturers.

Mike Sargent
Chairman, VAM Education and Training Working Group
LGC
October 2000

Acknowledgements

The authors would like to thank the following people who looked at various drafts of this handbook. Their advice, criticism and ideas have made this book a better product.

Members of the VAM Education and Training Working Group, in particular:

Dr Chris Amodio	Quality Assurance Agency for Higher Education
Professor Joe Connor	Medway Sciences
Dr Hywel Evans	University of Plymouth
Dr Colin Graham	University of Birmingham
Dr Sam Lalljie	Unilever
Dr Colin Osborne	Royal Society of Chemistry
Mr Douglas Squirrell	Consultant
Professor Ron Thomas	Consultant
Professor Alan Townshend	University of Hull

Our thanks also to Karen Rutland and Fraser Nicholson, LGC.

We are grateful for financial support from the Department of Trade and Industry as part of the National Measurement System Valid Analytical Measurement programme.

Figure 1 is redrawn from *Pure and Applied Chemistry*, 1990, **62**, 1193–1208 with the permission of the International Union of Pure and Applied Chemistry.

Introduction

This *Handbook* aims to explain terminology widely used, and sometimes misused, in analytical chemistry. It provides much more information than the definition of each term but it does not explain how to make measurements. Additionally, it does not attempt to provide comprehensive coverage of all terms concerned with chemistry, instrumentation or analytical science. The authors have addressed primarily those terms associated with the quality assurance, validation and reliability of analytical measurements. The *Handbook* attempts to place each term in context and put over concepts in a way which is useful to analysts in the laboratory, to students and their teachers, and to authors of scientific papers or books. This approach is particularly important because 'official' definitions produced by many international committees and organisations responsible for developing standards are frequently confusing. In a few cases the wording of these definitions completely obscures their meaning from anyone not already familiar with the terms.

The *Handbook* has two key features which are not usually found in scientific glossaries or other similar publications. Firstly, the contributors have attempted to explain not only the current, correct, usage of a term but also, where relevant, to indicate any differing usage which was common in the past or is still widely encountered. This is important because analysts frequently rely on books, guides or papers published many years ago and may well encounter usage of a term which directly conflicts with current practice. Similarly, newly-trained analysts work in the real world and may well encounter books or colleagues whose every-day use of a term differs from current recommended practice. Failure to understand what the original author or their colleague intended may lead to errors or even, in extreme cases, an accident. The second feature is the extensive use of practical examples which attempt to illustrate the use of a term in circumstances which the reader may well encounter in the laboratory, *e.g.* whilst speaking with a client, *etc*. These examples have been clearly indicated in the text so that the reader may quickly and easily refer to them. Alternatively, those readers who wish to do so will find it easy to follow the main text without being distracted by the examples.

The *Handbook* has been divided into six main parts of which the first four address the logical progression of an analytical measurement from obtaining a sample to reporting the result. Clearly there are many ways in which such a complex process can be divided but the authors believe that it is helpful to both students and practising analysts to address it in terms of:

1. **The Sample** – which includes all aspects of taking an original sample and handling it in the laboratory.
2. **The Analytical Method** – which describes a variety of terms associated with the measured quantities and measurement procedures together with parameters used to quantify fitness for purpose.
3. **Reference Materials, Chemical Standards and Reagents** – which covers a wide variety of terms associated with standards and reagents used during an analysis, including calibration solutions prepared by the analyst.
4. **The Analysis** – which addresses terms encountered whilst undertaking the analysis, including different types of analysis and sources of error.

The two remaining parts of the *Handbook* cover more general terms, which do not form part of the analysis itself but nevertheless represent critical aspects of obtaining and reporting reliable data. Part 5, **Maintaining Quality**, includes a wide variety of terms encountered in quality systems and documentation whilst Part 6, **Statistical Terms And Data Handling**, brings together definitions of those 'mathematical' terms most likely to be encountered by analysts in their everyday work.

It is not suggested that the *Handbook* be read from cover to cover as an introductory text. The thematic approach should, however, enable readers to obtain a feel for the variety of terms encountered with any given aspect of an analytical measurement and the ways in which they relate to each other. For those who miss the more traditional alphabetical listing of definitions, the Index provides easy access to both the definitions and the accompanying examples.

Contents

Table of Figures

1 The Sample

In the text the word sample is used in two ways. The **sample** referred to in Part 1 is the portion of material submitted for chemical analysis, *e.g.* a 10 mL sample of olive oil removed from a one litre bottle of oil for the purpose of chemical analysis. When statistical procedures are applied to the results of analytical measurements (Part 6) the term sample refers to the segment of all the possible results, *i.e.* the population that is being used to calculate the statistic. Ten measurements are used to calculate the sample standard deviation (see random errors, Section 6.2); the ten results are only a portion of the infinite number of possible measurements.

Terms used in describing samples and sampling procedures are many and varied. The text below aims to describe those in common usage. It is clear that there is potential for confusion. It should therefore be a golden rule for any analytical chemist to define clearly the terms used in the context of the particular problem being addressed so that any procedure can be understood and repeated.[1-3]

1.1 Sampling

Sampling	**Sampling schedule**
Sampling plan	**Sampling scheme**
Sampling programme	**Sample variability**

One of the first considerations in any analytical work is what information is required from the analysis. Fundamental to this is the sample and the procedure by which it is obtained, *i.e.* **sampling**. Consideration needs to be given to whether the sample is part of a larger whole, how it should be removed, and how it should be stored and processed prior to the analytical measurement.

Samples are taken from an infinite variety of sources and the sampling procedure has to be designed accordingly. There should be a requirement to hold discussions between the customer, the analyst, the sampling officer *etc.* to reach agreement on what is a representative sample.

Example: *1.1*

Sampling

Sampling a day's production of shampoo, already divided into several thousand individual containers, is very different from sampling water from a reservoir. Various means of removing samples are available (sampling method).

The method of sampling selected should be appropriate for the analysis. Sampling a tank of stratified liquid by dipping an empty sample bottle in from the top may be inappropriate, as this would discriminate with respect to the top layer. This method of sampling may, however, be suitable if the top layer was the object of the analysis.

With the exception of portable on-site testing kits, analytical equipment is rarely sited adjacent to the sampling point. Consideration should be given to the need for sample storage and transportation. For unstable samples it may be appropriate to stabilise the sample in some way before analysis can be carried out, *e.g.* this may involve chilling the sample.

Example: *1.2*

Sampling programme, sampling plan, sampling schedule, sampling scheme

The use of a sampling programme is common practice in the operation of a major chemical manufacturing plant. Analysis is carried out to ensure the plant is running efficiently and that feedstock and products are within the specified limits. Samples of the feeds and production streams are, therefore, taken regularly. The sampling plan is formulated for each analyte or group of analytes and combined into a sampling schedule or sampling scheme where the sampling point, sampling times, and frequency of sampling are defined.

The following require different sampling plans:

If feedstock is taken from a well-mixed supply tank for periods of days without the tank being refilled it is clearly unnecessary to analyse the feedstock each hour.

If the composition of production streams is known to vary slowly on changing the operating parameters of the plant then the sampling should reflect the fact that, even if operational changes are being made, analyses will not be required at a higher frequency than the time constant of the process.

Samples are taken from designated and carefully chosen sample points to meet the needs of the process. Therefore a sample taken to ensure product quality may be taken prior to the product storage tank so, if problems are identified, the material can be diverted to an off-specification tank for reprocessing.

Sampling of materials which are at non-ambient conditions also require special considerations – this is clearly the case with compressed gases and liquids where removing the sample under different conditions can result in a change to the sample.

Generally, the sampling process used must reflect a detailed consideration of the requirements of the analysis and the material being sampled.

A **sampling plan** needs to be established which involves decisions as to when, where, and how the sample should be taken, and whether the process of sampling should be a one-off or whether it should be repeated and how often. Safety implications of sampling procedures, sample preservation and transportation should be considered in any sampling plan. When there is a regular requirement for analysis the sampling plan is referred to as a **sampling scheme** or a **sampling schedule**.

The term **sampling programme** is commonly used to describe a combination of procedures where several related sampling schemes are combined.

Example: *1.3*

Sample variability

Samples taken from a water supply reservoir may vary with the time of day and the position from which they are taken. These factors would have to be considered in order to determine an overall analytical result for the reservoir.

Samples can vary with time, or with location, and the range of samples conceivably attainable is referred to as the **sample variability**.

1.2 Types of Sample

Aggregate sample	**Judgmental sample**
Batch	**Lot**
Bulk sample	**Random sample**
Composite sample	**Representative sample**
Gross sample	**Retained sample**
Heterogeneous sample	**Selective sample**
Homogeneous sample	

The term **bulk sample**, sometimes used interchangeably with **gross sample**, is an amount of material from which further sub-divisions are to be made. The **lot** is the amount of bulk material from which a particular sample is taken and can in some cases be identical to a **batch**, which is material produced or taken at the same time.

Figure 1 shows the relationship between the various operations in a sampling scheme up to sample taken for analysis. The terms used provide an alternative to those used here.

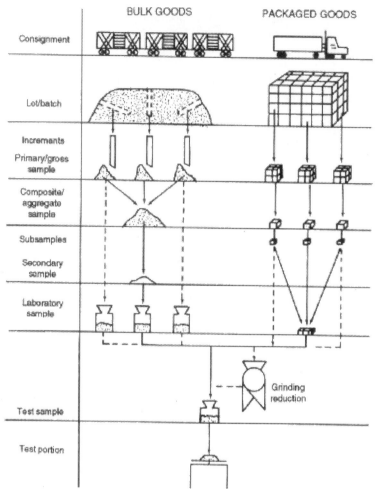

Figure 1 *Schematic sampling operations*

A **homogeneous sample** is one for which there is no evidence of variation throughout its extent.

Example:	*1.4*

Homogeneous sample

A drinking water sample would be homogenous for sodium ions.

Conversely, a **heterogeneous sample** is one that varies throughout its extent; any sub-samples taken from it may be expected to vary in composition.

Example: *1.5*

Heterogeneous sample

A soil sample made up of stones, clay, organic matter, *etc.* would be heterogeneous for iron(II) ions.

A **composite sample**, sometimes referred to as an **aggregate sample**, is one that is made up of several individual samples. Thus, a heterogeneous material may be sampled several times and these samples combined to produce a composite sample that is aimed to have the overall properties of the material.

A **representative sample** is one that is regarded as having identical composition to the larger bulk or batch from which it has been taken. Results obtained from a representative sample can thus be regarded as also applying to the bulk of material from which the representative sample was taken (see also Integrity, Section 1.3).

A **selective sample** is one which has been chosen to demonstrate a particular aspect of the material from which it is taken, *e.g.* a sample of water taken from immediately downstream of an industrial outfall to assess the worst possible situation with regard to potential pollution. Some analysts refer to **judgmental samples** as those chosen to illustrate a particular property of the material under consideration.

A **random sample** is chosen from a bulk material at random, *i.e.* so that any part of the bulk has an equal chance of selection, and without any particular reference to the properties of the bulk.

A **retained sample** is a sample taken for future reference purposes, *e.g.* for use in analysis where the result is the subject of a dispute.

1.3   Sample Handling

Aliquot	**Sample accountability**
Integrity	**Sub-sample**
Laboratory sample	**Test portion**
Primary sample	**Test sample**

The integrity of a sample should be maintained during the process of sample handling. The **integrity** of a material is described as the closeness of the state of the material to its original form. In the case of a certified reference material (see Section 3.1) it would be the closeness of the state of the material to that described on the certificate. During the sample handling and preparation stages of an analysis, the procedures should not change the information contained within the sample.

Example: *1.6*

Integrity

Along with the identity of a sample, its integrity is also important in forensic cases. An athlete has supplied a urine sample for drug testing. If the sample is contaminated at any stage then the integrity of the original sample (as described on the sample label) has been destroyed and the analytical results are worthless and cannot be used in evidence.

A term related to integrity is **sample accountability**, which is the procedure by which the appropriate documentation regarding the sampling process and subsequent actions prior to analysis is established and maintained. The sample will have a unique identity and each stage of handling the sample will be documented so that its whole 'life cycle' is traceable. Sample accountability is thus part of the audit trail (see Section 5.1) of a sample.

A **primary sample** is the sample taken from the lot or batch; in most cases a portion of this sample will be taken and known as a **sub-sample**. The sub-sample or part of the sub-sample is used to provide the laboratory sample.

A **laboratory sample** is the material sent to or received by the laboratory.[4] The **test sample**, which is sometimes the same as the laboratory sample, provides the source of material used to carry out the analysis. In many cases the laboratory sample has to undergo some type of sample preparation before it is ready for analysis.

Example: *1.7*

Laboratory sample → Test sample
100 g of wet sediment was submitted to the laboratory for analysis. The laboratory sample was dried at 30 °C for 24 hours and then homogenised to provide the test sample, a portion (test portion) of which was used for analysis.

A **test portion** is the amount of the test sample that is used for a particular determination.

An **aliquot** in this context is a part of a laboratory sample used in the analytical determination: the term is used commonly for the individual portions taken to repeat a single analysis.

Example: *1.8*

Aliquot

To perform the analysis, the analyst takes two 10 mL aliquots from the 100 mL test (sample) solution.

1.4 Sample Preparation

Clean-up   **Extraction**
Digestion   **Pre-concentration**
Dissolution   **Pre-treatment**

Samples, as taken, are often unsuitable for direct analytical measurement. These samples will require some pre-treatment. Sample **pre-treatment** is a term used to encompass a variety of sample preparation procedures, including pre-concentration, clean up, extraction, dissolution, digestion and homogenisation.

Samples are **pre-concentrated** in order to allow detection to be achieved. This is required when the concentration of analyte in the sample is too low or when as a result of pre-treatment the concentration in the resulting solution is too low for detection. Care is needed in any pre-concentration procedure to ensure that it does not lead to contamination from, for example, the equipment or chemicals used.

Example: *1.9*

Pre-concentration

A water sample requiring heavy metal determination may be evaporated to a smaller volume prior to using atomic absorption spectrometry.

Extraction procedures involve the removal of the analyte from a solid or liquid sample so that the analytical requirements can be achieved more satisfactorily than before, *e.g.* liquid–liquid extraction or solid-phase extraction.

Dissolution is the conversion of a solid or gaseous material into a solution. **Digestion** is the process of bringing a material into solution using, *e.g.* a solvent, often with the application of heat. Both are processes of preparing a sample for analysis. In doing so a heterogeneous sample is made homogeneous. Dissolution is commonly used in the pharmaceutical industry for preparing tablets for analysis.

Clean-up procedures involve the separation of analyte(s) from other sample components that might otherwise make the analysis more difficult. Filtration, sublimation and centrifugation are techniques that can be used for sample clean-up.

1.5   Sample Blanks

Blanks as part of sampling are described in this section whereas blanks used in the analytical procedure are covered in Section 5.4.

Field blank
Sample blank
Transport blank

Sample blanks are samples that are used to establish the effect of the sample matrix or to ensure that the matrix is not affecting the analysis. They approximate to the analytical samples except that they do not contain any measurable concentration of the analyte under investigation.

Sample blanks used for specific purposes may have more specific titles, such as transport blanks and field blanks. **Transport blanks** are taken through exactly the same journey as the analysis samples apart from the inclusion of the analyte. **Field blanks** are taken through the sampling process and are subject to all the possible sources of contamination without having the **analyte** included.

Example: *1.10*

Transport blank, field blank

Organic vapours in air can be sampled by pulling air, by means of a pump, through a tube containing charcoal. A transport blank would be a charcoal tube taken to the sampling site but never removed from the sampling case. A field blank would be a charcoal tube taken to the site, the ends of the charcoal tube broken and then connected to an air sampling pump without the pump being switched on.

---

2   The Analytical Method

Analytical methods are used to qualitatively and quantitatively establish the composition of a material. An analyst is often faced with answering questions of the type:

- Can this unknown substance x be identified?
- How much is there of x in y (*e.g.* lead in water, pesticides in fish, oil in a shale) or is there a suitable way of estimating the amount of substance x present? Is it necessary to take account of other chemicals present in y when performing the analysis?
- If the analysis is repeated using identical samples will the results be identical?
- If two different people use the same analytical method to analyse identical samples, will they both find the same amount of x in y?
- If an alternative analytical method is used, will it give the same result for identical samples?

This Part deals with the language associated with analytical methods.

2.1   Analysis

Analysis is the measurement of the amount of substance x present in the sample using an analytical method; everything else in the sample is the matrix. Some general names that one can give to x and the process of analysis are:

Analyte
Matrix
Measurand
Test

When an analysis is performed the amount of a chemical component present in a sample is identified and measured. The chemical entity being investigated (qualitatively or quantitatively) is called the **analyte**. There may be components, other than x, present in the sample. All the 'other' components are called the

matrix. The process of analysing the sample to determine the amount of the analyte in the matrix is called a **test**.

Example: *2.1*

Analysis, analyte, matrix, test

Consider the analysis of paint for lead content. The lead is the analyte; the paint is the matrix; and the process of analysing the paint sample for lead is the test.

Frequently, the amount of analyte present in a sample is determined indirectly, *e.g.* by measuring something that is proportional to the amount of analyte. Therefore, a more accurate way to describe what has been measured in an analytical method (test) is the term **measurand**. Measurand is defined as a "particular quantity subject to measurement".[5]

Example: *2.2*

Measurand

An analysis of paint for lead content is based on the absorption or emission of radiation by lead atoms. The lead atoms are the measurand.

2.2 Methodology

This section is about what type of test is appropriate to determine the amount of *x* in *y* (*e.g.* lead in water). Also dealt with is the question of why some types of tests are called methods, whereas others are called procedures, techniques, *etc.*

The names given to various types of analytical tests, developed over the years, are not used consistently and can cause much misunderstanding among analysts. This confusion has been reduced through the work of standardisation authorities that have tightened up on the use of terms and their meaning.

Terms associated with tests include:

Analytical technique	**Protocol**
End determination	**Reference method**
Measurement procedure	**Regulatory method**
Method	**Standard method**
National standards body	**Standard operating procedure**
Primary method	**Validated method**
Procedure	

When asked to determine the amount of a substance in a matrix, an analyst will use knowledge and experience to select an appropriate method to perform the analysis. This leads to a simple but effective way of defining a method. The **method** is the whole process involved in analysing a sample.

Alternatively a method may be defined as: 'a defined application of a particular technique or techniques to obtain an analytical result'. The approved definition for a method of measurement is, "A logical sequence of operations, described generically, used in the performance of measurements".[5] The method will encompass a range of procedures or steps such as sample preparation, sample extraction and measurement technique. Also it may be appropriate to have a procedure on how to obtain a representative, homogenous sample. A **procedure** is a set of instructions that are sufficient for the analyst to carry out the work.

Example: 2.3

Procedure

A procedure might be for the, 'Preparation of an extract of chlorophenoxy acidic herbicides from water using C18 solid phase cartridge'. The procedure will contain a set of instructions, for the analyst to follow, on how to obtain the extract solution. Another procedure will be required to explain how to analyse the extract solution using gas chromatography.

However, procedure can refer to other types of documentation (see Section 5.3).

Measurement procedure is the documentation that describes the set of operations used in performing the measurement; it is always part of the analytical method.

Example: 2.4

Measurement procedure

An analyst is determining the concentration of chloride ions in water using an electrochemical method. The measurement procedure is the documentation for the part of the method that gives instructions on how to make the measurements using a chloride electrode.

At some point during an analytical method, the analyst will have to identify and/or quantify the chemical entities that are being measured. The analyst will need to select an appropriate **analytical technique** in order to carry out this task.

An **analytical technique** is defined as a generic application of a scientific principle that can provide information about a sample.

Example: 2.5

Analytical techniques

(a) Electrochemistry – *e.g.* use of a pH electrode to measure the pH of a sample.
(b) Gas chromatography–mass spectrometry – used to separate, identify and quantify mixtures of organic compounds in a sample extract.

The **analytical technique** is normally the last practical stage of a method and this is why it is also known as the **end determination**. The end determination often involves the detection and quantitation of an analyte using an instrument.

Example: 2.6

End determination

An example is the determination of mercury in foodstuffs using cold vapour atomic absorption spectroscopy. The organic matter in foodstuffs is destroyed by wet oxidation. The mercury in solution is then reduced to the metallic state and released as a vapour in a stream of air. The quantity of mercury vapour in the air stream is measured by cold vapour atomic absorption spectroscopy. This part off the method is called the end determination. Often, as in this example, the end determination is equivalent to the analytical technique for the method.

It is a common assumption that if different analytical laboratories use the same procedure on identical samples they obtain virtually identical results. This assumption can prove to be incorrect due to different interpretations of how to perform the method. It has even been found that if two analysts working in the same laboratory are given aliquots of the same sample to analyse using the same method, frequently they will obtain different results.

A common reason for large variability of results is a lack of detail in the procedure which leads to different interpretations of what needs to be done. A solution to this particular problem is to produce a comprehensive document called a standard operating procedure.

A **standard operating procedure** is "a detailed written instruction to achieve uniformity in the performance of a specific function".[6] There will be standard operating procedures to cover all aspects of the analysis, *e.g.* registration and storage of the sample, documentation and reporting of results, training of staff, *etc.* Laboratories compliant with **Good Laboratory Practice** (see Section 5.2) are required to produce documentation for analytical methods in the form of standard operating procedures.

In addition, standard operating procedures are often written for well established analyses which are performed on a regular basis in a laboratory, often by more than one operator (see Section 5.3).

Another term used is **protocol**. Protocol has more than one meaning. In terms of a hierarchy of analytical methods, it is the highest level. In a protocol every step of the method is precisely defined and must be followed without exception, to the level of detail provided. A protocol is typically recognised or endorsed by a standards body.

A **protocol** is also used to describe a set of guidelines appropriate to a particular analytical area or generic group of operations. Examples include, a protocol to ensure a safe working environment, a protocol for the best way to choose the right method, a protocol for the conduct of collaborative tests, or a protocol for the best way to keep an analytical step or steps free from contamination.

A **standard method** consists of a set of instructions about how to carry out a method issued by a national standards body. A **national standards body** is a standards body recognised at the national level – in the United Kingdom it is the British Standards Institution (BSI). The standard method will have undergone a process of public consultation with the aim of producing a clear, concise and complete method. The precision of the method will have been checked, *e.g.* by collaborative study (see Section 5.6). To prevent a proliferation of standard methods, a standard method may be issued by more than one body, *e.g.* BSI, ISO and CEN.

Example: 2.7

Standard methods

The use of standard methods can save time and money. A manufacturer of ceramic ware needs to export a product to several different European countries. In the past, different laboratories for each individual country that the goods were imported into would have tested the ceramic-ware (for metal release). Now the manufacturer can pay one laboratory to use an approved standard method (BS ISO CEN 6748 – 'Limits of metal release from ceramic ware and vitreous enamel ware').

When the national government legislates to make a standard method mandatory it becomes a **regulatory method**. This method then permits trace-ability for this type of analysis.

When using a method, an analyst wants to know that the method is

performing satisfactorily. A **primary method** is one having the highest metrological qualities, whose operations can be completely described and understood,
and for which a complete uncertainty statement can be written down in terms of
SI units.[7] An analyst who can achieve the published level of performance of the
method, using known samples, can have confidence in the test results obtained
using the primary method.

Example: *2.8*

Primary method

Consider an analytical method involving the titration of hydrochloric acid
with anhydrous sodium carbonate to determine the concentration of the
acid. The measurements made are mass (weighing out a chemical to make up
a solution of known concentration) and volume (dispensing liquids with
pipettes and burettes). The reaction between the two chemicals is based on
amount of substance – one mole of sodium carbonate reacts with two moles
of hydrochloric acid – and the mass of a mole is known (*e.g.* the formula
weight in grams of one mole of sodium carbonate is 105.99). All the
measurements are based on either length or mass and are traceable to SI
units, so the method is a primary method.

Note that in clinical areas, primary method has a different meaning; in this
context it is a screening test (see Section 4.1).

A **reference method** is a method of known and demonstrated accuracy. It has
undergone a collaborative study using materials (*e.g.* certified reference material) where the amount of the analyte present in the material and its uncertainty
are known. This provides the analyst with performance data both on the
precision and on the accuracy of the method.

In practice, the number of **primary** and **reference methods** is limited. Normally
an analyst will use methods that are not based on length/mass/time, and where
there are no suitable reference methods available. The number of certified
reference materials available to evaluate the performance of a method is also
very limited. This makes it more difficult for an analytical laboratory to
demonstrate to its clients that the methods used will produce results that are
sufficiently accurate and precise to allow the client to reach a valid judgement.
An accepted way of proceeding is to carry out method performance checks,
which lead to a **validated method**.

Parameters usually examined in the validation process are: limit of detection, limit of quantitation, bias, precision, selectivity, linearity, range and
ruggedness. Limits will be set for each relevant parameter, and if these are
achieved during the performance tests the results are documented and the
method is said to be fit for purpose and now is a **validated method** for a given
scope (see Section 2.3).

Example: 2.9

Validated method

A forensic laboratory analyses blood samples from drivers suspected of driving under the influence of alcohol. The legal blood alcohol level for driving is 80 mg of alcohol in 100 mL of blood (80 mg%). The method used to analyse the blood samples will have been thoroughly validated at this concentration, to determine its accuracy and precision so as to ensure that innocent drivers are not found guilty, *etc.* If a lower limit of 50 mg% is introduced then the method will have to be validated at this lower concentration.

2.3 Performance Characteristics

Some of the common terms used in defining the performance of a method are:

Accuracy	**Repeatability**
Bias	**Repeatability conditions**
Conventional true value	**Repeatability limit**
Detection limit	**Repeatability standard deviation**
Discriminate	**Reproducibility**
Discrimination threshold	**Reproducibility conditions**
Instrument parameters	**Reproducibility limit**
Intermediate precision	**Robustness**
Limit of detection	**Ruggedness**
Limit of quantitation	**Scope**
Linearity	**Selectivity**
Precision	**Sensitivity**
Range	**Signal-to-noise ratio**
Recovery	

For the results of an analysis to be of use to a client, the analytical measurements need to be fit for the purpose for which they were intended. Key measures for determining the appropriateness of the analytical method used by a laboratory include bias, accuracy and precision.

If it is found that a method consistently produces higher (or lower) results for the analysis of a substance of known composition, the method is said to be biased.

The **bias** of a measurement result is defined as a consistent difference between the measured value and the true value. In mathematical terms it is the systematic component of measurement error.

The bias is a measure of the difference between a mean value of a set of results and a stated value (*e.g.* Certified Reference Material, Section 3.1).

Accuracy is often used instead of bias and trueness. It can be seen from Figure 2 that it involves bias and precision.

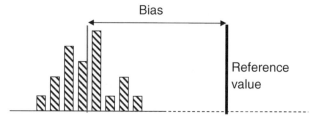

Figure 2 *Bias of measurement results*

Example: *2.10*

Bias

A glass pipette is used to dispense several measured aliquots of an aqueous sample solution. The temperature of the sample solution is 22 °C but the glass pipette is calibrated for solutions at 20 °C. There will be a bias in all the measured volumes due to the relationship between volume and temperature.

Accuracy is a term that is used loosely in daily conversation. It has particular meanings in analytical measurement. For example, under the current ISO definition, accuracy is a property of a result and comprises bias and precision.

Accuracy (of a measuring instrument): "Ability of a measuring instrument to give responses close to a true value" (see Section 6.2).[8]

Accuracy (of a measurement): The "closeness of the agreement between the result of a measurement and a true value of the measurand".[8]

Example: *2.11*

Accuracy (of a measurement)

The true value (see Section 6.2) of the amount of analyte in the sample matrix is not normally known. However, in some cases, certified reference materials (CRM) (see Section 3.1) are available. The analyst can perform an analysis on the CRM. The difference between the measured mean value and the stated value is a measure of the accuracy of the method (also known as trueness, bias).

The true value (see Section 6.2) of a measurand should be known to determine the bias of a method. The true value is the value that would be obtained by a perfect measurement. However, true values are by nature indeterminate: it is never actually possible to eliminate all measurement uncertainties. For some quantities a conventional true value is used.

Conventional true value is defined as a "value attributed to a particular quantity and accepted, sometimes by convention, as having an uncertainty appropriate for a given purpose".[5]

Example: 2.12

Conventional true value

A conventional true value that all analysts are aware of is the number of atoms in a mole – 6.022×10^{23}. However, it is known accurately and if required 6.0221367(36) can be used.

Precision is defined as "the closeness of agreement between independent test results obtained under stipulated conditions".[9] The precision of an analytical method is evaluated by making repeat independent measurements on identical samples to determine the spread of results.

Precision can be defined in mathematical terms as the random component of the measurement error of the method, so it does not relate to the true value. A useful statistical measure for precision is standard deviation or coefficient of variation (relative standard deviation). A large numerical value indicates that the process has a large component of random error, *i.e.* it is imprecise (see Section 6.2).

It is important to convey clearly what precision relates to in terms of the analysis.

Example: 2.13

Precision

A solution is divided into ten portions. Each solution is measured. The spread of the results obtained is a measure of the precision of the measurement. When this is quoted it should also state the conditions under which the measurements were made, *e.g.* all the values were obtained by the same person on the same day using a particular instrument.

Precision results obtained under some well defined conditions are normally expressed as **repeatability, reproducibility** or **intermediate precision**.

Repeatability (of results of measurements) is the closeness of agreement between the results of successive measurements of the same measurand on identical test portions carried out under defined conditions. Conditions will

---

Example: 2.14

Precision in terms of analysis

An analytical method is used to determine polychlorinated biphenyls in transformer oil by solvent extraction and gas chromatography. The precision is quoted as $\pm 10\%$ (2 standard deviations). However, this value needs to be qualified to be of use, *i.e.* does it refer to:

(a) the instrumental stage of the method, *i.e.* the gas chromatographic analysis of the extract solution;
(b) the analysis of the test portion of the sample, *i.e.* the extraction and gas chromatographic analysis stages;
(c) the complete analytical procedure including random errors which occur during the sampling of the transformer oil.

---

include the same operator, same apparatus, same laboratory, and short interval of time between analyses. The conditions under which these measurements were performed are known as the **repeatability conditions**.

The results of the repeatability experiments can be used to calculate a standard deviation, which is often called the **repeatability standard deviation**. This value is useful in helping the analyst determine a **repeatability limit** – "The value less than or equal to which the absolute difference between two test results

---

Example: 2.15

Calculation and use of repeatability limit

An analyst needs to calculate the value at which point there is a 95% probability that there is a significant difference between two single test results.

The analyst performs 10 measurements under repeatability conditions and has determined the repeatability standard deviation to be ± 0.12.

The repeatability limit is calculated using the following formula:[10]

$$r = t \times \sqrt{2} \times s_r$$

where

r = repeatability limit
t = Student t, two tailed value (95% for 9 degrees of freedom in this example)
s_r = repeatability standard deviation

A repeatability limit of 0.27 is obtained using the data in this example. The analyst can state that, at the 95% confidence level, two results differing by 0.27 are statistically significantly different.

obtained under repeatability conditions may be expected to lie with a probability of 95%".[9]

It is often necessary to estimate the spread of results obtained in different laboratories. A laboratory needs to show that the results from an analytical method are reproducible. **Reproducibility** is similar to repeatability except that the analyses are carried out on identical samples under **reproducibility conditions** (*e.g.* different operator, different apparatus, different laboratory, long interval of time). **Reproducibility limit** is similar to repeatability limit except that the results are obtained under reproducibility conditions.

Intermediate precision is the within laboratory variation over a long period of time. The standard deviation will be intermediate in value between that obtained under repeatability and that obtained under reproducibility conditions for similar samples using the same method.

Remember a method may perform well in terms of repeatability and reproducibility of results on identical samples, but this does not necessarily mean that the results are accurate.

In analytical chemistry the reason for a biased result may be due to analyte being lost or gained before the end determination stage. The bias is then usually reported as a recovery factor. The **recovery** is the factor or percentage of the total amount of a substance obtained after one or more manipulative stages of a method. Recoveries are commonly quoted both for an entire method and for specific parts of a method such as an extraction or a clean-up step. This will require a sample of known concentration being put through the analytical method or the particular step being investigated.

A key parameter in determining whether a method is suitable for use in trace analysis relates to the smallest amount of an analyte that can be detected. There are several terms used to describe this. Each has a specific meaning and the analyst needs to make clear the meaning of the terms used.

Many analytical methods use an instrumental technique for the final stage of the analytical procedure, *i.e.* the **end determination** (see Section 2.2). The detector of the instrument will have a **detection limit** which is the point where, with a stated probability, one can be confident that the signal due to the measurand can be distinguished from the instrumental background signal.

Example: *2.16*

Instrumental background signal

Even when no sample is being analysed, the instrument will be generating a background signal (*e.g.* a voltage) along with some noise caused by: (a) voltage changes in the electronic components; (b) static electricity.

   Note: The **signal-to-noise ratio** is taken "...as the magnitude of the background signal divided by the standard deviation of the background signal".[11] The limit of detection for an instrument improves if the signal-to-noise ratio can be increased. Instrument manufacturers should quote how they determine the signal-to-noise ratio.

In practice an instrumental detection limit is of limited use because in analytical chemistry it is rare that no other procedural steps are involved. Normally a limit of detection for the whole analytical method is required. The terminology used in this area is confusing. In general, limit of detection and detection limit are synonymous. The detection limit will encompass factors such as (a) sample matrix effects; (b) loss of the analyte during sample preparation; *etc.* The **detection limit** for the analytical procedure is defined as "The minimum single result which, with a stated probability, can be distinguished from a suitable blank value".[8]

There is sometimes a requirement to determine the value of the limit of detection for the analytical method.

The **limit of detection**, is derived from the smallest measure, x_{LOD}, that can be detected with reasonable certainty for a given analytical procedure. The value of x_{LOD} is given by the equation:

$$x_{LOD} = \bar{x}_{bl} + k s_{bl}$$

where $\bar{x}_{bl}$ is the mean of the blank determinations,
k is the numerical factor chosen according to the confidence level desired and number of blank readings taken,
s_{bl} is the sample standard deviation of the blank determinations.

Further information on determining an appropriate limit of detection and a suitable numerical factor (k) can be found in reference 12 or for more detailed treatment reference 13.

Example: *2.17*

Example of determining a detection limit for a method

An analyst needs to determine the limit of detection for an analytical method used to determine the concentration of lead(II) ions in water by inductively coupled plasma-mass spectrometry. The analyst performs several replicate analyses on independent sample blanks and obtains a mean concentration of $0.01~\mu g~L^{-1}$ and a sample standard deviation of $0.03~\mu g~L^{-1}$.

   The limit of detection for the analytical method is calculated to be $0.01 + (3 \times 0.03) \rightarrow 0.1~\mu g~L^{-1}$ using a numerical factor of 3.

Another way that has been used to describe the **limit of detection** of an analytical method is, "the detection limit of an individual analytical procedure is the lowest amount of an analyte in a sample which can be detected but not necessarily quantified as an exact value".[14]

In practice it is not easy to determine the detection limit. It is more common to determine the lowest concentration at which the analytical procedure performs reliably. This is known as the limit of quantitation or is sometimes referred to as the quantitation limit or limit of determination.

The **limit of quantitation** is the lowest concentration of an analyte that can be determined with acceptable uncertainty (see Section 2.4). The limit of quantitation is the point at which the results from a method become sufficiently unreliable to make quantification suspect.

Example: *2.18*

Example of determining limit of quantitation

Determining an acceptable limit of quantitation is a matter of judgement.

For example, analysis of blank sample solutions of lead(II) ions gave a sample standard deviation (s_0) of 0.1 μg L^{-1}. For most trace level analysis the level of quantitation is taken to be 10 s_0, therefore the limit of quantitation in this example is calculated to be 1 μg L^{-1} with an acceptable uncertainty of 10%.

The **range** of a method is "the interval between the upper and the lower concentration of the analyte in the sample for which it has been determined that the method is applicable".[15] However, another use of the word is to state that the **range** is the difference between the greatest and smallest values of a set of measurements. The latter meaning is commonly used by statisticians. Example 2.19 gives an example of a method where the linearity has been checked from 0.02 to 0.30 mg L^{-1}.

The **scope** of the method for measuring nitrite concentration shown in Example 2.19 would be covered in terms of concentrations and type of matrix for which the method applies, *i.e.* 0.05 to 0.25 mg L^{-1} in aqueous solution.

A term frequently linked with range is **linearity**. This is the ability of the method to give a response which is proportional to the concentration of analyte.

Sensitivity is defined as "the change in the response of a measuring instrument divided by the corresponding change in the stimulus".[5] In analysis the stimulus is often the amount of measurand present in the sample under test. Therefore, with reference to Example 2.19, Figure 3, sensitivity is the slope of the calibration curve. Note: In some applications, particularly clinical and medical, the term sensitivity is often associated with the lower limit of applicability (*e.g.* limit of detection) of a method.

When referring to an instrument, the term **discriminate** is used, *i.e.* the ability of a measuring instrument to respond to small changes in analyte concentration. This leads to the concept of **discrimination threshold**, which is defined as the "largest change in a stimulus that produces no detectable change in the response of a measuring instrument, ...".[5]

Laboratories need to know how reliable their analytical methods are in

> *Example:*   *2.19*
>
> **Range**
>
> A spectrophotometric method is used to measure the concentration of nitrite ions in water. The spectrophotometer response is shown to be linear from 0.02 to 0.30 mg L^{-1}. Therefore, the method has a working range from 0.05 mg L^{-1} to 0.25 mg L^{-1} nitrite concentration as shown in Figure 3.
>
>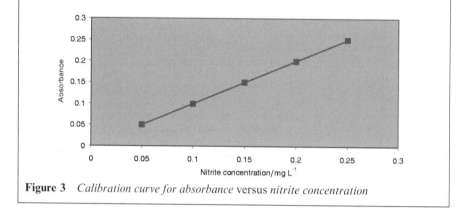
>
> **Figure 3**   *Calibration curve for absorbance* versus *nitrite concentration*

routine use. Participation in **Proficiency Testing schemes** (see Section 5.6) has shown that not all analytical methods perform satisfactorily over time or in different locations. One way to test the performance of a method over time is to determine the **ruggedness (robustness)** of a method. The **ruggedness** of a method is tested by allowing the parameters to change by a known amount (*e.g.* temperature of extraction) and measuring the effect on the result. Ruggedness gives an indication of how closely specified parameters need to be controlled, *e.g.* temperature to $\pm 1\,°C$, pH to ± 0.1 unit, *etc.*

> *Example:*   *2.20*
>
> **Testing a method for ruggedness**
>
> Checking a method for ruggedness will highlight the main parameters that need to be carefully controlled. For example, in the determination of cadmium release from ceramic ware, acid is added to the item of ceramic ware and left for 24 hours. During this period, the time, temperature and acid concentration will determine the amount of metal leached. Experiments show that temperature is a key parameter (results change by *ca.* 5% per °C) and it needs to be strictly controlled, *e.g.* to within $\pm 1\,°C$. Time is far less critical (result changes by *ca.* 0.3% per hour) and therefore needs to be less strictly controlled, *e.g.* ± 0.5 hour.

As part of the ruggedness testing, the analyst will need to ensure that any instrument is working correctly and that all variables are under control.

An **instrument parameter** is an aspect of the instrument which, if altered, may change the output from the instrument. An instrument parameter is best described using a specific example.

Examples: 2.21

Instrument parameters

When using an ultraviolet/visible spectrophotometer, instrument parameters will include wavelength and slit width. If the wavelength is changed then the measured absorbance of the sample solution is likely to change.

Analytical methods need to be selective.

Example: 2.22

Selectivity of an analytical method

An analytical method is used to estimate the amount of gold in sea water samples. The results of the analysis will be used to decide if it is economic to extract gold from sea water. Therefore, it is important that the concentration determined by the analytical method is only due to the gold ions in solution and is neither enhanced nor suppressed by other components of the matrix.

Selectivity of an analytical method is 'the extent to which, in a given procedure, other substances affect the determination of a substance according to a given procedure'.[8] When using an analytical method, consideration needs to be given to the effect that (a) impurities, (b) degradants and (c) the matrix might have on the ability of the method to discriminate between these components and the analyte.

An ideal analytical method has the ability to discriminate between the analyte and everything else present in the matrix. This ultimate **selectivity** is often termed **specificity**.

2.4   Fitness for Purpose

An analyst needs to select a suitable method for carrying out an analysis. The method needs to be validated and its measurement uncertainty determined.

Terms associated with this topic include:

Combined standard uncertainty	**Measurement uncertainty (Uncertainty of measurement)**
Coverage factor	
Error	**Method validation**
Expanded uncertainty	**Standard uncertainty**
	Uncertainty component

All laboratories need to demonstrate that their methods are fit for the purpose for which they are intended. The process of obtaining data to demonstrate this fitness for purpose and the resulting documented evidence is called **method validation**. The purpose of method validation is to determine that one or more of the performance characteristics of the method is satisfactory so that the analytical results obtained will be fit for their intended purpose.

Table 1 *Performance parameters*

Performance parameter	Procedure that could be used
Limit of detection	Analyse blank sample matrix
Limit of quantitation	Analyse spiked (low level) blank sample matrix
Bias estimate	Use CRM (see Section 3.1) or spiked samples
Linearity range	Plot response of a set of solutions of known concentration against their concentration
Selectivity	Analyse CRM Analyse spiked sample matrix
Precision (1) Repeatability (2) Reproducibility	 Repeat analysis of either a CRM or a homogenous sample Several laboratories carry out repeat analyses of either a CRM or a homogenous sample
Ruggedness	Identify components of the procedure that require special control

Table 1 contains examples of what needs to be checked in the process of method validation and the procedures that could be followed. This list is not meant to be comprehensive.

All results will be subject to **random errors** and there may be **systematic errors** (see Section 6.2). This means there will be some uncertainty in the final result.

The **uncertainty of measurement** is defined as a "parameter, associated with the result of a measurement, that characterises the dispersion of the values that could reasonably be attributed to the measurand".[5] It is also known as measurement uncertainty. A simpler way of stating what measurement uncertainty means is to say that **measurement uncertainty** is the range of values within which the value of the quantity being measured is expected to lie with a stated level of confidence. It is not the same as an error, because to estimate an error the true value must be known. **Error** is the difference between a 'true value' and the result.

The laboratory often has a need to know the overall **measurement uncertainty** for an analytical method because the result is being used to evaluate compliance with a statutory limit. Quoting a measurement uncertainty with the result enables the laboratory to give the client more useful information for judging compliance.

Example: 2.23

Use of measurement uncertainty

A water sample is analysed to check that it complies with the maximum admissible concentration of nitrate in water (50 mg L^{-1}). The analytical result can be reported in one of two ways:

(a) 49 mg L^{-1} nitrate in water
(b) 49 mg L^{-1} ± 4 mg L^{-1} nitrate in water (providing a level of confidence of 95%)

If a client was given result (a) they would think that they had complied with the regulation. Result (b) shows that because of the measurement uncertainty of the method the result is somewhere between 45 and 53 mg L^{-1} so it is possible that the client may have breached the regulation.

In practice the **measurement uncertainty** for a particular analysis can arise from many sources. Each source will contribute to the overall uncertainty of the measurement and so each source is considered to have its own **uncertainty component**.

Example: 2.24

Uncertainties in different formats

Uncertainties are often obtained from different sources and are in different formats:

A manufacturer of Grade A 1000 mL volumetric flasks stamps on the side of the flask the following information: 1000 ± 0.4 mL. Therefore, a 1000 mL volumetric flask will have a volume between 999.6 mL and 1000.4 mL.

A manufacturer of a four-figure balance states that the balance reading is within ±0.1 mg with 95% confidence.

An analyst measures the concentration of fat in margarine on five identical samples. The sample standard deviation for the set of results is a measure of the uncertainty for the analytical method.

None of these uncertainties are in the same format but each piece of information can be used to calculate the corresponding standard uncertainty.

To be able to combine measurement uncertainties from the various components in an analytical method, they all need to be in the same form, *i.e.* as a standard deviation. When expressed as a standard deviation, an uncertainty component is known as a **standard uncertainty** (see Section 6.3).

The analyst normally requires the total uncertainty of the result. This is termed the **combined standard uncertainty** and is made up of all the individual standard uncertainty components.

Example: 2.25

Combined standard uncertainty

For quantities that are added together the method of combining the individual standard uncertainty components is to use the root sum of squares:

$$u_c = \sqrt{(u_1{}^2) + (u_2{}^2) + \ldots}$$

where u_c is the combined standard uncertainty and u_1, u_2, *etc.* are the individual standard uncertainties (*NB* uncertainties are in the form of a standard deviation).

For most purposes in analytical chemistry an **expanded uncertainty** (U) should be used when reporting a result, *i.e.* $x \pm U$. U is calculated using the following equation:

$$U = k \times u_c$$

where k is the coverage factor and u_c is the combined standard uncertainty.

The **expanded uncertainty** provides the range within which the value of the measurand is believed to be for a particular level of confidence (see Section 6.2).

Example: 2.26

Use of expanded uncertainty and coverage factor

A report to a client contains a statement as follows:

The concentration of nitrate ions in the solution is,

$$10 \pm 0.5 \text{ mg L}^{-1} *$$

The client then knows that the true value lies in the range 9.5 to 10.5 mg L^{-1} and can base any judgement on this fact.

* The reported uncertainty is an expanded uncertainty calculated using a coverage factor of 2 which gives a level of confidence of 95%.

The **coverage factor** chosen depends on the level of confidence required. For most purposes $k = 2$ is used, where it is assumed that the distribution is normal and the level of confidence is 95%.

3 Reference Materials and Chemical Standards

Of the two terms, reference materials and standards, the latter has so many meanings that care needs to be taken in its use. For example, is the analyst talking about the standard of the work (*i.e.* the analysis), the standard solutions prepared or the standard method used. Reference material has a much less ambiguous meaning in the analytical laboratory.

3.1 Traceability and Reference Materials

Reference materials provide analysts with a common reference point, *i.e.* when analysing a reference material there is some property of the material that can be referred to which is well defined or characterised. Some of the terms associated with reference materials are:

Certified reference material (CRM)　　**Natural matrix reference material**
In-house reference material　　　　　　**Reference material**
Matrix reference material　　　　　　　**Traceability**

The ISO definition of a **Reference material** (RM) is "a material or substance one or more of whose property values are sufficiently homogeneous and well established to be used for the calibration of an apparatus, the assessment of a measurement method, or for assigning values to materials".[16]

The *raison d'être* of reference materials is to provide **traceability**. Ultimately and ideally all chemical analyses will be traceable to an SI unit, the amount of substance (mole) or mass (kg). In this way, agreement between different analysts, laboratories, and countries is enhanced.

There are a few analytical techniques, *e.g.* weighing, where it is possible to directly show **traceability** to an established value. For most cases, *e.g.* to verify an analytical procedure, one has to resort to the use of reference materials. Analysis of a reference material means that the analyst can check the result against an established and agreed number.

The overall hierarchy for the use of reference materials in the laboratory should be as follows:

1. Certified Reference Materials (including matrix reference materials)
2. Reference Materials

3. In-house Reference Materials
4. Spiked samples

The ISO definition of a **certified reference material** (CRM) is "Reference material, accompanied by a certificate, one or more of whose property values are certified by a procedure, which establishes its traceability to an accurate realisation of the unit in which the property values are expressed, and for which each certified value is accompanied by an uncertainty at a stated level of confidence".[16]

Example: *3.1*

Traceability using a certificate of calibration

Balances are calibrated using a standard set of weights that come with a certificate of calibration giving their mass and uncertainty of this value, *e.g.*:

Serial number: CC 100   Date of Issue: 24 November 1999

Nominal value	*Measured value*	*Uncertainty of measurement*
100 g	100.000 38 g	± 0.1 mg
1 g	1.000 011 g	± 0.02 mg

It is possible to eventually trace these values back to the mass of the standard kg. This chain of comparison of masses achieves the necessary traceability. It is important that the chain is unbroken and that there is a stated uncertainty at each link.

A lead nitrate certified reference material (CRM) could be a solution of that compound in, say, nitric acid at a designated purity and concentration that has been tested by sufficient methods for an organisation to put some numbers on a certificate. Alternatively, it may be lead nitrate present in a milk powder with a value determined by several techniques from a collaborative testing scheme, again with a certificate. The latter is an example of a **matrix reference material**. The certificate will give details of the concentration and the uncertainty in the measured value at, *e.g.* a 95% level of confidence.

When developing methods for analysing samples that are not simply a single substance (*e.g.* metals in skim milk powder) it is good practice to use a matrix reference material (see previous example) of the same/similar composition as the sample under investigation. A **matrix reference material** is a homogeneous material, *e.g.* milk powder, containing known concentrations of one or more analytes. Similarly a **natural matrix reference material** is a natural material (*e.g.* one found in the environment) that has been homogenised and for which the concentration of one or more analytes has been established.

Example: 3.2

Certified reference material

There is a certified reference material from BCR (Bureau Communitaire de Reférence) – Catalogue number: CRM 063R, Skim milk powder (natural). The analyst who purchases this CRM will receive 50 g of the material and a certificate.* Part of the information on the certificate is:

Element	Certified value	Uncertainty	Units
Calcium	13.49	0.10	mg g^{-1}
Copper	0.602	0.019	μg g^{-1}
Lead	18.5	2.7	μg g^{-1}

The analyst would use this **CRM** to check the performance of an analytical method; *e.g.* determination of lead in milk.

* A detailed technical report about sample preparation, certification, *etc.* is also supplied.

An **in-house reference material** is an RM prepared by an organisation solely for its own internal purposes. The values for this material may have been obtained by comparison with a CRM.

Example: 3.3

In-house reference material

This term might be used to describe a metal alloy sample known to be inherently stable, tested for homogeneity as part of routine production quality control and analysed by both the laboratory's routine method (say, XRF) and an alternative method (say, dissolution and measurement of the solution by ICP–OES).

When no suitable reference material is available it is possible to add a known amount of the pure analyte to a matrix of the same composition as the samples (or those containing a very low natural amount of analyte). This is called spiking (see Section 5.4).

3.2 Chemical Standards

The term standard is used in many laboratories and has more than one meaning. ISO have included primary and secondary standards under terms related to reference materials.[16]

Some of the many terms associated with standards are:

Analytical standard   **Secondary standard**
Calibration standard   **Stock solution**
Drift correction standard   **Working standard**
Primary standard

In chemistry the term *standard* has traditionally been used to describe a solution of a pure substance and of known concentration.

Now, with the advent of reference materials, the term standard is too general to be used alone and it is, therefore, normal to specify the type of standard, *e.g.* analytical, calibration, reference material, *etc.* Use of the term standard(s) is, therefore, only appropriate in the most general of contexts.

The general function of all types of **analytical standard** is either to calibrate parts of the method (almost always the measurement step) or to check the performance of the whole method, *i.e.* the work-up of the sample as well as the measurement step. An **analytical standard** is the general term used to describe a solution or matrix containing a known amount of analyte, which will be used to check the performance of a method/instrument.

A **calibration standard** is a pure compound of accurately known concentration used to calibrate an instrument.

Example: *3.4*

Calibration standard

An analyst needs to calibrate an atomic absorption spectrophotometer used for the determination of lead in milk samples. The analyst can purchase a certified reference material (CRM) solution of lead in nitric acid. This CRM is used to prepare a set of calibration standard solutions of known concentration of lead. These calibration standards, along with a suitable blank material (see Section 5.4), are used to calibrate the instrument.

In many methods an analyst has to prepare a set of **calibration standard** solutions to calibrate an instrument. The solutions are prepared by dilution of a concentrated analytical standard. The concentrated analytical standard is known as the **stock solution**. The solutions used to perform the calibration are described as **working standards**.

There are analytical methods where it is impracticable to carry out a full calibration after every set of measurements. The analyst will need to demonstrate that the calibration has not significantly changed before carrying out another set of measurements. The analyst can prepare drift correction standards to carry out this check. A **drift correction standard** is a standard solution of known concentration used to monitor the calibration of an instrument.

Example: 3.5

Drift correction standard

The analyst uses ICP–OES (inductively coupled plasma, optical emission spectroscopy) to measure twenty different metal ions in solution. To fully calibrate the instrument requires the preparation and measurement of 100 individual calibration standards (five point calibration per element). It would be impracticable for an analyst to calibrate the instrument daily. The instrument is calibrated at regular intervals (say fortnightly) by the analyst. In the intervening time, the calibration for each metal ion is checked by the use of a set of drift correction standard solutions. Minor corrections can then be made to the calibration to allow for day-to-day drift.

Two other terms commonly used in the laboratory are primary and secondary standards.

Primary standard is defined as a "standard that is designated or widely acknowledged as having the highest metrological qualities and whose value is accepted without reference to other standards of the same quantity".[5]

Example: 3.6

Primary standard

Arguably, the ultimate primary (chemical) standard is silver (often referred to as 'five-nines' grade). This is available commercially with a stated purity of 99.9995% Ag.[17]

A primary standard is used to establish a secondary standard. A **secondary standard** is defined as "a standard whose value is assigned by comparison with a primary standard of the same quantity".[5] It is often the secondary standard that the analyst will use.

3.3 Properties of Chemicals

This section is about the chemicals used to carry out an analysis and the terms given relate to answering questions of the type:

- Is the chemical pure?
- Is the chemical stable over time?
- Has the identity of the chemical been established?

The meaning of the following terms are explained in this section:

Authentic reference spectra   **Labelling**
Authenticity   **Purity**
Expiry date   **Reagent**
Grade

Purity is defined as the fraction of the named material present in the stated chemical form. When using standards (including reference materials) it is obviously important to know what their composition is in terms of their purity.

Example: *3.7*

Purity

An analyst purchases a certified reference material (CRM). The CRM is accompanied by supporting documentation. Included in this documentation will be details of both (a) the estimated/established purity, *e.g.* 99.99% and (b) the impurities with their concentration.

All analytical laboratories handle numerous samples and have to prepare countless standards and solutions. The labelling of each sample and solution is crucial if there is to be confidence in the analytical results. **Labelling** is the process of assigning a unique identifier to each sample, standard, *etc.*

Example: *3.8*

Labelling

In forensic cases it is important to ensure that all samples are labelled correctly and that there is a chain of custody (see Section 5.3) back to the point from where the sample was taken. Consider the case of two athletes who have urine samples taken for analysis for the presence of illegal drugs. One of the tests is positive and the other negative. The analyst who gives evidence in court must be certain beyond reasonable doubt that the positive result from the urine sample can be connected to the correct athlete. This can be achieved if the samples and supporting documentation are clearly and accurately labelled.

When **labelling** a freshly prepared standard solution, the label should include the following information:

Analyte:	Lead nitrate
Concentration:	0.1 mol L^{-1}
Analyst who prepared the solution:	A Chemist
Preparation Date:	29 December 2000
Expiry Date:	29 December 2001

Other information which is useful to record on the label: the name of the chemical manufacturer, the grade of the material, the lot or batch number and the purity of the material.

Example: *3.9*

Expiry date

When using any analytical standard, the analyst needs to know that it has not significantly deteriorated. Most standards and reference materials are supplied with a date estimated by the manufacturer after which they should not be used. Expiry date is the date after which the chemical should not be used. There is often a difference in expiry date depending on the physical state of the chemical.

When determining an **expiry date**, the analyst should take into account the frequency of use of the standard. The number of times the bottle is opened or aliquots removed and the temperature at which it is stored will have an impact on the integrity of the standard. Once the integrity of the standard has been significantly altered then the standard has to be disposed of and a new standard made.

The degree to which the information written on the label is 'true at the time of receipt' is the **authenticity** of the standard. The **authenticity** of the standard is best determined on initial receipt of the standard by comparison with other or previous standards known by the analyst to be traceable and conform to the specifications on the label.

A standard (or sample) may also include some additives that have been added to stabilise the solution. This should also be detailed on the label. Additives should not significantly affect the authenticity of the standard. A typical example is where a small amount of nitric acid is added to a solution of an analytical standard to keep metal ions in solution during storage.

Methods of analysis normally require other chemicals besides standards and samples. **Reagent** is defined as "...a test substance that is added to a

---

Example: *3.10*

Reagent (reactant)

An analyst is measuring the concentration of nitrite ions in water. The method used involves the preparation of a solution of sulfanilamide. This solution is added to the samples and standards to produce a red complex that is measured spectrophotometrically. The sulphanilamide solution is known as a reagent because it caused a reaction to occur.

system in order to bring about a reaction or to see whether a reaction occurs".[8] Reactants are similar to reagents except they are consumed in the reaction process. Therefore an indicator can be classed as a reagent but not a reactant.

Like standards, reagents will be classified according to a grade. The **grade** will indicate the purity of the material.

Example: *3.11*

Grade

Common manufacturers' grades include: AnalaR (AR), Aristar, Specpure, Spectrometric, SLR (Specified Laboratory Reagent), GPR (General Purpose Reagent) and Technical.

Potassium thiocyanate can be obtained in two grades from one manufacturer:

SLR grade for 'general laboratory usage' > 98% purity
AR grade 'certified reagent for analysis' > 99.5% purity

Different analyses will require different grades of chemicals, *e.g.* an analytical method used to measure very low levels of analyte will require a specially high purity grade of reagent such as Spectrometric or Specpure. The nature of the impurities is also given on a label and this is important.

National and international compendial [pharmacopoeial] authorities publish **authenticated reference spectra**. **Authentic infrared [IR] reference spectra** are required for proof of identity of 'official' substances by direct comparison with observed IR spectra obtained with the test substance examined in a prescribed manner.

The **authentic spectra** are generated on instruments of good, but not necessarily the highest, quality and may, for convenience of publication, be electronically reduced. An expert panel responsible to the competent compendial authority authenticates each spectrum.

Authentic IR reference spectra covering the range 2000–500 cm^{-1} are published in conjunction with monographs for substances of the British Pharmacopoeia (*BPC, London*). They are also supplied by the European Pharmacopoeia (*EDQM, Strasbourg*) for a more limited number of official substances, where the cost, or toxicity, of a suitable chemical Reference Substance, or restrictions on its supply by post, favour identification by comparison with an official Reference Spectrum. The International Pharmacopoeia (*WHO Collaborating Centre for International CRS, Stockholm*) will supply on request paper or electronic copies of an authentic Reference Spectrum covering the full mid-IR range, *i.e.* 4000–600 cm^{-1}.

The advantages of issuing authentic Reference Spectra include:

- major savings in the cost of creation, maintenance and distribution of official Reference Materials;
- avoid collaborative exercises for the validation of such materials;
- no problem of deterioration or special storage conditions;
- avoid any problems in the distribution of drugs subject to international control.

Some disadvantages of wholly relying on authentic Reference Spectra for identification of official drugs include:

- no access to corresponding chemical reference material;
- inability to compare a Reference Spectrum obtained with different types of instrument, *e.g.* to compare spectra from interferometric [FT-IR], with another such as grating dispersive transmittance spectrometer;
- no opportunity for direct validation of locally generated spectra loaded on to a data station.

4   The Analysis

This part deals with the language associated with carrying out the analysis.

4.1   Types of Analysis

Analysis is about finding the amount of the substance of interest (analyte) in the sample.

Characterisation	**Qualitative**
Confirmatory	**Quantitative**
False negative	**Screening**
False positive	**Semi-micro**
Macro	**Semi-quantitative**
Major	**Trace**
Micro	**Ultra-micro**
Minor	**Ultra-trace**

An analyst normally tests a sample to identify the components in the sample and/or determine the amount of some or all of these components.

Example: *4.1*

Test

The test may be to identify an unknown white powder, *e.g.* is it sugar or heroin? The test may be to determine the amount of Ag per cm^2 of photographic paper.

Qualitative analysis is the process of determining if a particular analyte is present in a sample or the identity of the sample.

Often the identification of the analyte in the matrix provides insufficient information. The analyst, as well as identifying an analyte, will be required to measure the amount of analyte present in the sample.

Example: 4.2

Qualitative

Identification of an unknown substance, *e.g.* analysing an insulation board to determine if it contains asbestos.

Quantitative analysis is the process of measuring the amount of the analyte that is present in the sample matrix.

Example: 4.3

Quantitative

Determination of the concentration of nitrate ions, mg L^{-1} in a water sample, or the amount of fat, mg kg^{-1}, in margarine.

It is not always necessary to identify an analytical species. Sometimes it is sufficient to identify and/or quantify a specific property or attribute of the sample material, *e.g.* materials stable below 60 °C, non-flammable. This is known as **characterisation** of the sample.

Example: 4.4

Characterisation of organic compounds

The analyst needs to select all carbonyl compounds from a set of available liquids. The characterisation test will need to be specific to carbonyl groups. *NB* The exact identity of the materials will not be obtained.

It may be sufficient to obtain an approximate measurement of the concentration of analyte present in the sample. This is known as **semi-quantitative** analysis. A **semi-quantitative** method provides an estimate of analyte concentration, perhaps to within an order of magnitude or sometimes better.

Sometimes it is necessary to analyse a large number of samples in a short period of time to identify samples that give a positive result. Screening methods are often used in this situation. The **screening** method used must be relatively inexpensive, capable of rapid analysis of samples and reliable enough not to produce large numbers of false positive results or false negative results. In general a more thorough (but invariably slower and more expensive) analytical method will be applied to the samples for which a positive result is

---

Example: 4.5

Semi-quantitative

Semi-quantitative analysis is common in situations requiring on-site analysis. An analyst is trying to identify and quantify the amount of airborne pollution present in an underground car park in order to determine if there is a health problem. Pollutants will be mainly due to emissions from car engines (carbon monoxide, hydrocarbons, benzene, nitrogen oxides, *etc.*) and will vary significantly depending on the number and type of cars using the car park. There is a range of chemical indicator tubes (tubes containing indicators that change colour in the presence of specific chemicals) that can be used to identify and quantify the pollutants. The results obtained will give an indication of the concentrations of the pollutants but it will only be an approximation.

---

obtained. The samples that give a false positive result are identified and eliminated by this subsequent analysis.

Example: 4.6

Screening

Screening tests are performed on suspected drugs of abuse. Suspect materials are tested using a TLC method to identify amphetamines and opiates. A more thorough analytical method is used to determine the exact analytes present in the samples that gave positive TLC results.

False positive results are those results that show a particular analyte is present in a sample when in fact there is none present. Similarly, **false negative** results show absence of the analyte when it is actually present.

When there is doubt about an analytical result, it may be necessary to carry out a further test. This is termed a **confirmatory** analysis if it involves a different technique based on an independent principle.

When performing an analysis of the sample, the choice of method will be influenced by the quantity of sample available. The following terms refer to scale of working: macro, semi-micro, micro and ultra-micro.

Macro analysis is the term used when referring to situations where there is a significant amount of sample used in the test.

Similarly:

- **semi-micro** analysis refers to analysis of a small amount of sample;
- **micro** analysis refers to analysis of a very small amount of sample;
- **ultra-micro** analysis refers to analysis of a minute amount of sample.

However, in practice one finds that in different laboratories, the actual quantity of sample being tested (*e.g.* for macro analysis) will vary enormously.

The situation is equally ambiguous when discussing the concentration ranges measured. Four terms are commonly used to describe the concentration range of an analyte measured in a sample:

- **major**;
- **minor**;
- **trace**;
- **ultra-trace**.

However, there is a lack of agreement as to what these concentration ranges are. Suffice it to say that major > minor > trace > ultra-trace when talking about concentration levels.

Trace is also used in another context when taking about a very small quantity of sample.

Example: *4.7*

Trace (in a different context)

A forensic method involves the identification of blood group from a trace of blood on the victim's clothing.

Here, the word trace is being used to describe the quantity of blood on the victim's clothing.

4.2   Sources of Error

This section deals with some of the problems associated with analysing a sample. The general terms that occur in relation to problems associated with estimating the amount of *x* in *y* are:

Contamination	**Interference**
Cross contamination	**Interferent**
Homogeneity	**Speciation**
Inhomogeneity	**Stability**
Instability	

An analyst needs to know that the analytical procedure being used is measuring only the analyte from the sample. Before and during the analysis of the sample there is always a potential for contamination.

Contamination is defined as the unintentional introduction of analyte(s) or other species which were not present in the original sample and which may cause a bias in the determination.

Example: 4.8

Contamination

Sources of contamination might include:

- the working environment in which the analysis is performed;
- the chemical reagents used in the analysis;
- the glassware used to carry out the analysis;
- the sampling equipment used to take the sample.

Cross contamination is where some of the analyte present in one sample or standard inadvertently transfers to another sample.

Example: 4.9

Cross contamination

The results of an analysis for explosives are used as evidence to show that an individual has been handling explosives. A whole batch of evidence had to be rejected because it was found that the inside of a centrifuge used in the analysis had become contaminated with trace amounts of explosive from a broken centrifuge tube. It was not possible for the laboratory to prove beyond reasonable doubt that no cross contamination had occurred.

An **interferent** is the term used to describe any component in the sample that will affect the final measurement result. **Interference** is a systematic error (see Section 6.2) in the measurement caused by the presence of interferents in the sample.

Example: 4.10

Interferents and interference

An analyst is measuring the concentration of americium-241 (alpha emitter) in a soil sample. Most of the components in the soil will act as interferents as they physically block the alpha particles reaching the detector. The measured concentration of americium-241 in soil will be low due to the interference caused by the solid material in the sample. Therefore, the sample will require pre-treatment to remove the interferents before the americium-241 concentration can be measured.

In spectrophotometric measurements, for example, interference means any chemical species, which will emit or absorb energy at the same spectral wavelength as the analyte under investigation.

Interferents can affect the measurement in a positive or negative way. Even if no interference can be detected in a sample, the matrix may contain both positive and negative interferents, which cancel each other out. Whether a particular substance in a matrix is an interferent depends on the analytical technique being used.

Example: *4.11*

Interferent and the analytical technique used

The pH of potassium permanganate solution cannot be tested using universal indicator solution because the colour of the potassium permanganate solution interferes. However, the pH can be determined using a pH electrode since the colour of the solution does not affect the $H^+_{(aq)}$ ion concentration measurement.

Instability means a change in the chemical composition of a substance with time.

Example: *4.12*

Instability

Metal ions in water samples will often precipitate out if the sample is stored for an appreciable length of time. The sample is unstable. To keep the metal ions in solution the sample can be stabilised by the addition of a small amount of nitric acid at the time of sampling.

Stability is used when describing how a substance remains unchanged over a period of time.

Stability can also refer to the output of an instrument. The baseline from the

---

Example: *4.13*

Stability

'The standard solution, when prepared, should be stored at $4\,^\circ C$ in a refrigerator. The solution is stable for a month'.

---

gas chromatograph was checked for stability before injecting a sample solution, *i.e.* no drift or change in signal-to-noise ratio.

It is always a matter of concern when analysis of duplicate samples produces significantly different results. One of the key reasons for this is that the analyte is unevenly distributed through the sample matrix. Sample **inhomogeneity** is the term used to describe situations where the analyte is unevenly distributed through the sample matrix (see also heterogeneous, Section 1.2). Similarly, sample **homogeneity** is the term used to describe how uniformly the analyte is distributed through the sample.

When measuring the concentration of certain elements or components it is important to know which species are present.

Speciation is a term that refers to the exact form of an element or compound present in a sample. The concentration of an element or component measured will depend on the speciation and the method chosen to perform the analysis.

Example: *4.14*

Speciation

Mercury can be present in shellfish as mercury metal, inorganic mercury salts or as organomercury compounds.

The analyst has used an analytical method to determine the concentration of 'mercury' in a sample of shellfish. The analytical method chosen might measure total mercury (the metal mercury and all its compounds) or it might only measure certain mercury species (*e.g.* mercury(II) salts, organomercury compounds). Therefore, the method chosen may have a significant effect on the mercury concentration found in the sample.

---

5 Managing Quality

Quality in an analytical laboratory, or in any other organisation, can only be maintained by ensuring that the pursuit of quality is fully integrated into all laboratory activities and practised by laboratory staff at all times. The arrangements for managing quality can be divided into two groups. Firstly the relatively fixed arrangements such as administrative procedures and secondly dynamic processes that vary according to the nature of samples or tests, *e.g.* including characterised samples, known as quality control samples, with test samples.[18–21]

5.1 Quality System

A **quality system** is a set of procedures and responsibilities which a company or organisation has in place to ensure that staff have the facilities and resources to carry out measurements that will satisfy their clients. The quality system is a combination of quality management, quality assurance and quality control (see Section 5.4). The terms related to these activities are:

Assessment	**Management review**
Assessor	**Quality assurance**
Audit schedule	**Quality audit**
Audit trail	**Quality manager**
External audit	**Quality manual**
Horizontal audit	**Technical manager**
Internal audit	**Vertical audit**

Quality assurance is the planned and systematic control, implemented in the quality system, which provides confidence that the analytical service and the data generated provide the quality required by the client.

Internal quality assurance provides confidence to the management whereas external quality assurance provides confidence to the client. They are planned activities designed to ensure that the quality control activities are being properly implemented.

The **quality manual** is the primary source of reference for the quality management system. It describes the policies and procedures the laboratory has to manage quality, or references where those procedures can be found. A quality manual is essential to meet the requirements of the ISO 9000 series of standards and the ISO 17025 standard (see Section 5.2).

Example: 5.1

Quality manual

A quality manual has to document those procedures and policies that are carried out in the laboratory that can affect an analysis. In addition details of the organisational relationships, responsibilities and authorities of all of the more senior staff and the internal auditors are described. The manual will probably include descriptions of the resources of the laboratory, examples of records used, calibration and audit schedules, and routines such as the periodic review of the quality system.

A vital requirement of any quality system is that all aspects must be audited, at least annually, by a person independent of the activity.

A **quality audit** is an inspection made on a selected area of the laboratory, or of the quality system, in order to obtain objective evidence of the extent to which the planned arrangements for achieving and maintaining quality are met, and indeed whether or not they are adequate. Audits must ensure that day-to-day operations comply with the requirements of the system. (Is the laboratory doing what it says it does?)

An **internal audit** is one undertaken by an auditor who is an employee of the laboratory, usually independent of the area being audited.

An **external audit** is one carried out by an auditor external to the laboratory. This may be a representative of an accreditation or certification body or possibly a client.

There are two different types of audit and these are described in Example 5.2.

Example: 5.2

Horizontal audit and vertical audit

A typical way to perform a quality audit is to select a group of samples that have been tested, follow their progress through each stage of the analytical process, including all aspects of the analysis from the receipt of the samples to the final report to the customer. This is called a **vertical audit**.

A **horizontal audit** examines one aspect of the analysis. For example, a horizontal audit might check the balance used to weigh samples for analysis. Is the balance calibration checked as required by the quality system? Has it been regularly serviced? Have the weights used to perform the balance calibration themselves been calibrated?

The process for vertical audit described in Example 5.2 is an example of an **audit trail**.

Internal and external audits will be examined, usually annually, to check that the quality system meets current needs and/or that the procedures are being implemented. This formal process is known as **management review**.

It is common practice for a senior member of staff in the laboratory, often known as the **quality manager**, to have the responsibility of ensuring that the quality system is kept up-to-date. He or she will also ensure that periodic and systematic reviews of the system are conducted, and that an **audit schedule**, a timetable of when audits are to be performed, is prepared. An audit may highlight instances when practice deviates from the documented procedure. The quality manager must ensure that these digressions, known as **non-compliances** or **non-conformities**, are rectified within a timescale agreed with the auditor.

The **quality manager** must have a direct line of communication to the head of the laboratory, who may sometimes be referred to as the **technical manager**. In a small laboratory it is quite acceptable, for both ISO 9000 and ISO 17025 purposes, for quality and technical management to be headed by the same person, provided the dual responsibilities do not cause a conflict of interest. However, for GLP (see Section 5.2) compliance, these functions must be clearly separated and fulfilled by different individuals.

An **assessment** is always an external examination carried out by a laboratory accreditation body such as UKAS (see Section 5.2). It includes all of the elements of an audit, *i.e.* an in depth check that the quality system as documented is implemented and in operation. In addition, it involves a technical assessment of the laboratory to confirm that staff are qualified and competent and that the procedures being used for testing are fit for the intended purpose and able to achieve the level of performance claimed by the laboratory.

A lead **assessor** (and/or assessment manager) will perform the assessment supported by independent technical assessors with the expertise to look at the specific areas of the laboratory's work that are being accredited.

5.2   Quality System Standards

Accreditation	**GLP (Good Laboratory Practice)**
Accreditation body	**GMP (Good Manufacturing Practice)**
Accreditation standard	**ISO 17025 standard**
Certificate of accreditation	**ISO 9000 series**
Certification	**NAMAS M10/M11**
Certification body	**Surveillance**
Compliance	**UKAS (United Kingdom Accreditation**
CPA (Clinical Pathology Accreditation)	**Service)**

Terminology to indicate that the laboratory meets the requirements of a standard is used rather loosely. The preferred usage is:

- meeting the requirements of ISO 9000 series is Certification;
- satisfying ISO 17025 (a standard that specifically addresses technical competence) is Accreditation;
- strictly, compliance is only applied to GLP but tends to be used as a generic term.

An organisation may work to a number of different **quality system standards**. These standards are described in documents, which define the requirements that must be met.

In order to demonstrate its compliance with the standard a laboratory will normally apply to an external **accreditation body** or **certification body** for **accreditation** or **certification** against the standard. Certification involves confirmation that the quality management system in place in the laboratory is fully implemented and in compliance with the requirements of the standard. Accreditation, however, adds a further element and involves a peer review of the methods used by the laboratory to confirm that they are suitable for the purpose for which they are being offered. Certification involves checking that what is described in the quality manual is carried out, *i.e.* an **external audit** (see Section 5.1). On the other hand accreditation is offered on the basis of an **assessment** (see Section 5.1).

In addition to the initial inspection of the laboratory the accreditation or certification body will carry out **surveillance** visits, normally at annual intervals, to confirm that compliance is maintained. A **re-assessment** visit is carried out every four years. A laboratory may also seek to extend its scope of accreditation at any time by, *e.g.* seeking accreditation for a new technique.

The **ISO 9000 series** (International Organization for Standardization) are the quality management standards most commonly used by organisations manufacturing or supplying products or services in the UK and across the world. Laboratories may be certified against these standards but it is more usual, because of the additional scrutiny and hence additional credibility arising, for laboratories to seek accreditation against one of the standards specifically designed for laboratories. Laboratory **accreditation** is the "formal recognition that a testing laboratory is competent to carry out specific tests or specific types of tests".[22]

The most general laboratory **accreditation standard** is **ISO 17025**. This standard was published in December 1999 and is set to become the first truly international standard for laboratory accreditation. In this respect it replaces the UK's **NAMAS M10/M11** (National Accreditation of Measurement and Sampling) standard. In the UK the **United Kingdom Accreditation Service (UKAS)** acts as the accrediting body for the NAMAS standard and will continue to do so for ISO 17025 when the NAMAS standard is superseded.

Other standard protocols to which an analytical laboratory may need to seek **compliance** include **GLP** (Good Laboratory Practice), **GMP** (Good Manufacturing Practice) and **CPA** (Clinical Pathology Accreditation).

GLP is mandatory for analytical laboratories playing any part in toxicological study programmes carried out on new compounds for safety purposes. It is a subsection of GMP, which applies to laboratories performing quality control analysis as part of the manufacture of medical products and medical devices. In the UK, the Medicines Control Agency administers both GLP and GMP.

Clinical Pathology Accreditation as its name implies is a scheme specifically for the accreditation of pathology laboratories. The scheme is operated by Clinical Pathology Accreditation (UK) Ltd. which is a private company owned jointly by the Royal College of Pathologists, the Institute of Health Services Management, the Institute of Biomedical Science, the Association of Clinical Biochemists and the Independent Healthcare Association.

For ISO 9000 approval there are several certification bodies who register and certify companies meeting the requirements. In the UK such certification bodies must themselves be accredited by UKAS as competent to carry out certifications. For GLP, the GLP Monitoring Authority of the Medicines Control Agency issues a letter of compliance to an approved laboratory. This is restricted to laboratories carrying out safety studies as described above. However, any laboratory may state they work within the GLP guidelines. In the case of accreditation by UKAS, a **Certificate of Accreditation** is issued which relates to specific tests or calibrations. Laboratories will be included in UKAS's list of accredited laboratories, *i.e.* UKAS Directory of Accredited Laboratories.[23]

5.3 Documentation

There are several reasons why a quality system must be fully documented. Firstly it is a pre-requisite of most quality standards. Secondly, in most laboratories it would be impossible to accurately remember and hence communicate all of the analytical methodology and quality management procedure to staff. This would lead to the quality system becoming compromised due to staff turnover. Thirdly the process of audit (see Section 5.1) requires a precise definition of the planned quality system. This is provided by the documentation.

There are several terms used to describe the types of documentation used in a quality system in addition to the quality manual itself (see Section 5.1).

Archives	**Records**
Calibration schedule	**Sample register**
Chain of custody	**Standard operating procedure**
Controlled documents	**Study plan**
Methods documentation	**Test sheet**
Procedure	**Training record**
Protocol	**Unique sample identifier**

Terms for the different types and levels of documentation below the quality manual may vary. The terms in common use are described.

The term **procedure** is generally applied to the documentation describing activities to be carried out in the laboratory. Some laboratories may make a distinction between **procedures**, which describe activities other than the test methods themselves, and **methods documentation**; others will regard everything below the quality manual as a procedure.

Example: 5.3

Procedures

- how to record all samples coming into the laboratory;
- how test reports and certificates are to be issued;
- how to conduct internal audits and reviews;
- how to deal with complaints from customers;
- methods for calibration and performance checks on equipment;
- instructions for preparing, labelling and storing reagents;
- standardisation of titrimetric solutions;
- safe disposal of unwanted chemicals and samples;
- analytical methods.

The term **standard operating procedure (SOP)** is properly restricted to GLP operations (see Section 5.2 and below) where it has a precise definition within the terms of the standard. However, the term is widely used by laboratories not complying with GLP as a synonym for **procedures** to emphasise the importance of following them as standard practice (see Section 2.2).

The term **protocol** is usually applied to a set of procedures, covering sampling and sample preparation through to reporting the analytical results. The protocol is therefore a complete plan for conducting an analysis (see also Section 2.2).

In the context of a GLP study (see Section 5.2) the term **Study Plan** or **Study Protocol** refers to a protocol which defines in minute detail how work is to be conducted. The Study Plan will be formally agreed by the laboratory and the client and also vetted by the quality manager (see Section 5.1). In order to avoid having to repeat descriptions of common laboratory operations in every Study Plan the GLP laboratory will maintain a library of standard operating procedures which can be cross referenced, as appropriate, in any Study Plan.

All documents of the type referred to so far in Section 5.3 must be issued as **controlled documents**. This means that each copy is numbered and issued to a particular individual or location. A record is kept of each issue so that, when the document requires update, all copies in circulation can be updated and there is no danger of superseded versions of a document being followed in error. Controlled documents must neither be amended by unauthorised persons nor photocopied.

For all quality standards, staff working in a laboratory must be suitably trained. For laboratory workers a **training record** is a document used to record each analyst's training and the date when the analyst is deemed competent to carry out a particular task. It should be reviewed on a regular basis to identify training needs and skills that the analyst will require in the future.

In addition to descriptions of procedures to be followed, the quality system must provide for the creation and maintenance of **records** which show what has been done. This is necessary to enable the laboratory to replicate, if required, the analytical process that was performed for a particular sample. The process of audit is only possible if adequate records are maintained and the absence of adequate records will constitute a non-compliance with most quality management standards (see Section 5.2).

Example: *5.4*

Records

The types of records relating to the analysis include:

- the client's requirements;
- the methods used;
- analytical raw data, including all charts, spectra, print-outs, work sheets and note books;
- calibration and audit records;
- quality control checks including any proficiency testing;
- reference items employed;
- the identities of the equipment used and analysts;
- analytical reported results;
- interpretation and conclusions;
- correspondence;
- client report.

It is important, and for many analytical laboratories essential, to ensure that the progress of a sample throughout the analytical process can be tracked, and that there is an unbroken **chain of custody** as the sample passes from one area to another and from analyst to analyst.

Example: *5.5*

Chain of custody

Forensic laboratories use the system of chain of custody to demonstrate in court that the samples have, at all times, been held securely by authorised persons and that no tampering with the evidence has taken place in the laboratory. In addition, it is essential that one can demonstrate that the results of the analysis relate to the correct sample and that it can be traced back to the suspect or the site from which it originated.

When a sample arrives from a client it should immediately be assigned a **unique sample identifier**, often the next number in a defined sequence. This number should be attached to the sample and used by all the staff concerned. It must also be recorded either manually or electronically in a **sample register**.

Example: 5.6

Sample register

Typical information, which would need to be recorded in the sample register, includes:

- date of receipt;
- name of person making entry;
- name of analyst;
- client's description of the sample;
- client's identification and contact details;
- client's requirements;
- appearance, odour and condition of the sample on receipt;
- unique sample identifier;
- date of report.

In order to perform the analysis correctly, the analyst needs to be given the appropriate information about the analysis required. In large organisations the person who deals with the client may be remote from the analyst who carries out the work. The **test sheet** is where the client's request has been recorded with sufficient detail for the analyst to perform the analysis. With the advent of laboratory information management systems (**LIMS**), this may be done using a computer rather than a paper based system.

After the task is complete, including reporting to the client, all documents relating to it should be filed, preferably together, and placed in the laboratory's **archives**.

Example: 5.7

Archive

An archive is a secure filing area where records are held and where staff access is controlled. The storage medium may be paper, film or electronic. Most quality systems will specify a minimum time for records to be retained.

In any laboratory, a number of routine checks are performed to ensure continuing correct performance. Under a quality system, one very important routine is to check, at regular intervals, the calibration of equipment used. A **calibration schedule** is a timetable specifying the frequency of calibration of each item and the acceptance limits for the calibration measurements.

Example: *5.8*

Calibration schedule

A calibration schedule details the calibration of balances, volumetric glassware, automatic pipettes, thermometers, pH and conductivity meters, wavelength and photometric scales *etc.* The schedule consists of periodic external checks, employing a suitably accredited calibration service, supported by more regular in-house performance checks.

5.4 Quality Control

The term **quality control** (QC) is applied to procedures used to provide evidence of quality and confidence in the analytical results. It includes use of blanks, replication, analysing reference materials or other well-defined samples and participation in Proficiency Testing schemes. Several other features of analysis forming part of QC are control of reagents and instrumentation, equipment maintenance and calibration, and procedures for checking calculations and data transfer. It should be noted that what is referred to as quality assurance in the UK is known as quality control in Japan.

This section is about the types of materials used to establish whether analytical methods are performing reliably. The terms associated with this are:

Blank   **Reagent blank**
Check sample   **Replicate**
Duplicate   **Solvent blank**
Quality control sample   **Spiked sample**

A **quality control sample** is a sample of known composition, very similar in terms of matrix and analyte concentration to the samples to be analysed. It is analysed along with the samples.[7] The results are often plotted on a chart to monitor any significant drift in the performance of the analytical method – such a chart is known as a control chart (see Section 5.5).

A quality control sample, often known as a QC sample or **check sample** may be an in-house material prepared for this purpose or it may be from a previously analysed batch of samples.

A **spiked sample** is a sample prepared by adding a known quantity of analyte to a matrix which is close to or identical to that of the sample of interest. Spiked samples may be used in method validation experiments to help identify matrix effects and determine the recovery (see Section 2.3) of an analyte or the selectivity of the method.

Example: 5.9

Quality control sample

Brewing companies analyse their products on a regular basis to determine alcohol content. An aliquot of a homogenous beer, of known alcoholic strength, is used as a QC sample. This sample is analysed on a regular basis along with the samples, to monitor the performance of the laboratory method.

Example: 5.10

Spiked sample

A laboratory wishes to establish if its method for determining pesticide residues in fruit is performing satisfactorily. Invariably, most of the fruit samples will not contain any measurable concentration of pesticides. To obtain materials to check the performance of the method, in particular to gain an estimate of the bias, a laboratory could take a portion of unused sample extract that has previously been analysed and add a known amount of pesticide to the matrix. Analysis of this spiked sample will help demonstrate if all the pesticide is being recovered in the analysis or if there is significant interference from the sample matrix.

Replicate means repeat and may refer to samples or measurements or steps in the analytical procedure such as replicate injections onto a GC. If only two samples or measurements are involved then the term **duplicate** is used.[24] In practice the terms duplicate and replicate can be used in a variety of ways and it is important to describe in what context the terms are being used.

Example: 5.11

Duplicate (used in different contexts)

Duplicate analysis performed on the same sample – in this context the analysis can only be performed using tests which are non-destructive and which do not alter the sample in any manner.

Duplicate analysis performed on two test portions from the same sample – *i.e.* two test portions from the same sample are taken through the whole method.

Duplicate analysis of a sample extract solution – *i.e.* one sample has been taken through the analytical procedure up to the measurement step. At this point the extract solution is split into two portions and each portion measured.

Duplicate samples – two packets of Cornflakes from the same production batch.

Table 2   *Use of the term blank in analytical methods*

Term	Definition
Blank measurement	The measured value obtained when a specified component of a sample is not present during the measurement.[24]
Blank matrix	A matrix which does not contain the analyte above the limit of detection.[7] Ideally the matrix blank will have exactly the same composition as the sample matrix though in practice this is not always easy to achieve.
Blank sample	A sample whose analyte concentration is below the limit of detection of the analytical method being used.
Blank analysis	Describes a test procedure, which is carried out without the sample (see reagent blank). It can also mean analysis of the matrix material where no analyte of interest is present.

The term **blank** is used in many different contexts (see also Section 1.5). Some examples are shown in Table 2.

A **reagent blank** is a solution or mixture obtained by carrying out all the steps of the analytical procedure in the absence of a sample.[7] The reagent blank does not contain either the sample matrix or the analyte. It only contains reagents and solvents used for the sample preparation and the analytical method. In this context the reagent blank will be taken through the procedures in exactly the same way as the test sample.

A **solvent blank** is a portion of the solvent used as a blank probably only in part of the analytical procedure.

Examples: *5.12*

Reagent blank

Kjeldahl analysis consists of adding various reagents to the sample in a glass tube, performing a digestion and developing a coloured complex whose concentration is determined spectrophotometrically.

A reagent blank for this analysis consists of taking an empty tube and placing all reagents in the tube (no sample) as if performing the analysis.

Solvent blank

Polyaromatic hydrocarbons (PAHs) in water are determined by extracting the PAHs by shaking with hexane, concentrating the extract solution by controlled evaporation and then measuring the PAH concentration in the hexane extract by gas chromatography.

Hexane from the stock bottle put through the same procedure (*i.e.* concentrated and analysed by gas chromatography) would act as a solvent blank.

5.5 Quality Control Methodology

The terms associated with quality control methodology are:

Acceptable quality level   **Inspection level**
Action limits   **Tolerance limits**
Control chart   **Warning limits**

A **control chart** (quality control chart) is a graphical record of the results of the analysis of a quality control sample using a particular method. Monitoring these results over a period of time is one of the most useful ways of determining whether or not a method is in statistical control, *i.e.* it is performing in a consistent manner. It helps to indicate the reliability of the results.[7] There are many forms of control chart,[21] one of the most commonly used is the Shewhart Chart (Figure 4).

The data are plotted on a control chart in time sequence. This enables the analyst to readily observe changes in the measured value. The analyst can define warning and action limits on the chart to act as alarm bells when the system is going out of control. In Figure 4, it shows that all the results of the analysis of the QC samples are within the warning limits except for one result which is between the upper warning and upper action limit.

The **action limits** are the boundaries on the control chart at which the user must take action if any points exceed the defined limit. They are usually set at an appropriate confidence level, equivalent to the sample standard deviation value obtained from precision studies multiplied by a number. Warning limits are normally set at the 95% confidence level ($\equiv 2 \times$ sample standard deviation)

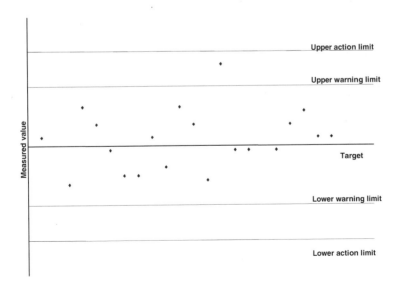

Figure 4   *Example of a Shewhart Chart*

and action limits at the 99.7% confidence level ($\equiv 3 \times$ sample standard deviation). The **warning limits** are the points on the control chart at which the user must be alerted if any points are exceeding these limits. If points continue in the sequence to exceed the warning limit then the user must take action.[21]

Tolerance limits (limiting values) are used in some quality control systems. For a test sample, the tolerance limit is the acceptable upper and/or lower limit of analyte concentration obtained using a specified analytical method. This means for a sample to be accepted as satisfactory, its analyte concentration must fall within prescribed limits.

When working in a production environment it is not practicable to analyse more than a few samples from a production batch. A **sampling scheme** (see Section 1.1) needs to be designed to decide on an appropriate number of samples to take and analysed to determine if the quality is satisfactory. **Inspection level** defines a relationship between batch size and sample size, *i.e.* numbers of units examined. Inspection levels are given in published International Standards.[25]

Example: *5.13*

Inspection level

A brewer has manufactured 100 000 cans of beer and needs to check that the alcohol content does not exceed 4%. The analyst can consult the relevant standard – BS 6001:1999 – to ascertain the number of cans that need to be selected in order to provide a sample representative of the whole batch.

Acceptable quality level is related to the quality required in the product. The **acceptable quality level** is the maximum percentage nonconformity that, for the purpose of the sampling inspection, can be considered satisfactory as a process average.[21]

The acceptable quality level has a particular significance in the design and use of acceptance sampling plans, *e.g.* ISO 2859 – 1.

Example: *5.14*

Acceptable quality level

The acceptable quality level (AQL) is given by the formula:

$$\text{AQL} = 100 \times \frac{\text{Number of nonconforming units}}{\text{Number of units examined}}$$

Therefore, an acceptable quality level of 0.010 would mean that one nonconformity in 10 000 is the limit of acceptance.

5.6   Performance

Laboratories need to monitor their performance on a regular basis to ensure methods are producing precise and accurate results. Some of the terms to do with this topic are:

Collaborative study	**Material certification study**
Consensus value	**Method performance study**
Interlaboratory study	**Proficiency test**
Laboratory performance study	**Z Score**

There is little benefit to be gained, when analysing identical samples using the same method, if the result of the analysis performed by one laboratory differs from the result from another laboratory. If the client does not know which laboratory to believe, the case might end up in court and the laboratories will probably blame each other rather than identify the cause of the problem. A laboratory needs evidence that the methods being used are performing correctly. There are many types of studies for evaluation of laboratories and their performance, these are listed below.

Interlaboratory study – A study in which several laboratories measure a quantity in one or more 'identical' portions of homogenous, stable materials under documented conditions, the results of which are compiled into a single document.[9]

Method performance study – An interlaboratory study in which all laboratories follow the same written method to measure a quantity in sets of identical test samples. The reported results are used to estimate the performance characteristics of the method.[9]

A **collaborative study** is used to evaluate proposed standard methods. It allows several laboratories to try out a proposed analytical method using identical samples. The results of the analyses are used to assess the performance characteristics of the method. The eventual aim of the collaborative study is to be able to produce a **standard method** (see Section 2.2) having a known **measurement uncertainty** (see Section 2.4) that can be used with confidence by any analytical laboratory competent in the field. A collaborative study is also used to provide information to the authors of the method so that ambiguous, wrong or misleading statements are corrected before the method is published. Methods that are likely to have undergone collaborative studies before being issued as standard methods include those issued by the:

- British Standards Institution (BSI)
- American Society for Testing and Materials (ASTM)
- International Organization for Standardization (ISO)
- Association of Official Analytical Chemists (AOAC)
- European Committee for Standardization (CEN)
- Analytical Methods Committee (AMC) of the Royal Society of Chemistry.

Material certification study – an interlaboratory study that assigns a reference value ('true value') to a quantity (concentration or property) in the test material, usually with a stated uncertainty.[9]

Laboratory performance study – An interlaboratory study that consists of one or more measurements by a group of laboratories on one or more homogeneous, stable, test samples by the method selected or routinely used by each laboratory.[9] The reported results are compared with the consensus value. Note, this is very similar to a Proficiency Testing scheme (see below).

Consensus value[26] can be defined as the mean of participants' results on a test material distributed in a proficiency testing scheme, after outliers have been handled either by elimination or by the use of robust statistics.

A **Proficiency Test** (PT)[27] is defined as 'the study of laboratory performance by means of ongoing interlaboratory test comparisons'. It is also known as an external quality assessment scheme, external laboratory performance check or external quality assurance (EQA). There are many such schemes run by independent external bodies for different analytes in a variety of matrices. Evidence in published papers shows that the performance of analytical laboratories improves as a result of participating in Proficiency Testing schemes and the between-laboratory precision can improve, sometimes dramatically. This is especially true in the early years of participation.

All proficiency testing schemes are based on the same principle. Laboratories are sent one or more samples to analyse. The results of the analytical measurements are returned to the organiser and then compared to the assigned value of the sample. The assigned value is determined by the organisers, it may be a consensus value from the results of the tests or a true gravimetric value. A statistical system is used to assign a score for the performance of each laboratory. The score the laboratory receives will depend on how close its result is to the assigned value.

One of the most widely used and simplest scoring systems used in PT schemes is the **Z** score. Z score is defined as 'a performance score recommended for use in proficiency testing schemes to evaluate the accuracy of the analytical results submitted by participating laboratories'.[28]

The **Z** score is calculated using Equation 1.[26]

$$\mathbf{Z} \text{ score} = \frac{\text{Laboratory Result} - \text{Assigned Value}}{\text{Target Standard Deviation}} \tag{1}$$

Z score of |2| or less is considered satisfactory.
Z score between |2| and |3| is a cause for concern.
Z score of over |3| is unsatisfactory.

6   Statistical Terms and Data Handling

6.1   Result

Calibration
Non-parametric statistical techniques
Outlier
Raw data
Reporting limit
Test result

The result of an analytical measurement is usually obtained by the application of appropriate statistical procedures to the **raw data** obtained in the laboratory, monitoring station *etc.* In many cases some of the data are derived from measurements on standards (materials of known composition), the remainder coming from test materials. In a **calibration** experiment several standard materials are examined using a carefully defined sequence of steps, involving appropriate sampling and preparation of the materials followed by an instrumental analysis protocol. One of the standards is a blank material, containing solvent, reagents *etc.* but no analyte. The results of these measurements are plotted as a calibration graph or analytical growth curve. Test materials ('unknowns') are then analysed using exactly the same sequence, and the results are used with interpolation on the calibration graph to yield a **test result** for each test material.

A problem that arises in many cases is that small raw data samples may contain one or more results which appear to be divergent from the remainder; the problem may arise in a calibration experiment or in a set of replicate measurements on a test material. Assuming that the cause of the divergence is not immediately apparent (*e.g.* as a result of a transcription error or instrument failure) it is then necessary to decide whether such an **outlier** can be rejected entirely from the raw data before data processing (*e.g.* logarithmic transformations) and the usual statistical methods are applied. The treatment of outliers is a major topic in statistics. Three distinct approaches are available.[29]

(1) Special **significance tests** (see Section 6.2) can be applied to establish whether the suspect measurement can be rejected at a particular con-

fidence level (see below). This approach suffers from the disadvantage that the tests generally assume a Gaussian or other specific error distribution in the measurements (an assumption that may not be valid), and are rather complex to apply if there are two or more suspect measurements at either or both ends of the range of results.

(2) **Non-parametric statistical techniques** (*i.e.* those that make minimal assumptions about the error distribution) can be used to handle the raw data. Such methods are generally resistant towards the effects of extreme values, often because they use the median (see Section 6.2) as a measure of central tendency or measure of location. Such methods have the further advantage of extreme simplicity of calculation in many cases, but while popular in the behavioural sciences they are less frequently used in the analytical sciences.

(3) **Robust statistical methods** (see Section 6.3), which may well represent the best current approach to the problems of suspect values, despite their requirement for iterative calculations.

It is very important to report the results of analyses in ways that are readily and unambiguously understood. The methods used to determine average values, standard deviations, confidence intervals, uncertainties *etc.* must be clearly defined. In some application areas, the concept of a **reporting limit** is useful: it is stated that the concentration of the analyte is not less than a particular value. Prudence indicates that this limit should be greater than the **limit of detection** (see Section 2.3). Use of an appropriate number of significant figures is also critical in reporting results, as this indicates the precision of the measurements.

6.2 Statistical Terms

Arithmetic mean	**Median (average)**
Bias	**Random errors**
Coefficient of variation	**Relative standard deviation**
Confidence intervals	**Robust variance**
Confidence levels	**Samples and populations**
Confidence limits	**Standard deviation of the mean (standard**
Degrees of freedom	**error of the mean)**
Error distributions	**Systematic errors**
Gross errors	**Variance**
Mean	

Statistical methods have to be applied to experimental results because of the occurrence of errors. The measurements made in practice are only a small **sample** from the potentially infinite **population** of measurements, and statistical methods (based on the laws of probability) are applied to give information about the population from information about the sample.

Example: 6.1

Samples and populations

If there are ten analysts working in a laboratory, they are the population of analysts in that laboratory. However, they are only a sample of the analysts in the whole world.

Three types of error are recognised:

(1) **Gross errors** are, if detected in time, so irredeemable that an experiment has to be repeated. They may give rise to **outliers** (see Section 6.1) in a set of results.
(2) **Systematic** (determinate) **errors** each cause all the results in a sample to be too high or too low: the sum of the systematic errors (with regard to their signs) produces an overall **bias** in an experimental result.

Example: 6.2

Systematic error

Sample solutions are prepared in volumetric flasks using distilled water at a temperature of 18 °C. The volumetric flask has been calibrated at a temperature of 20 °C. This will lead to a small systematic error in the results of analysis of the sample solutions because the volume will be different from the stated value.

Repetition of the measurements cannot by itself identify or estimate systematic errors. When such replicates are applied to the analysis of a standard (known concentration) material the bias of the analytical system can be assessed.

(3) **Random errors** cause the results of individual measurements to fall on either side of the mean value. They arise from many causes at each of the

---

Example: 6.3

Random errors

When preparing sample solutions in 100 mL volumetric flasks, the analyst needs to ensure that the correct amount of distilled water is added to each flask. A good technique is to ensure that the bottom of the meniscus is lined up with the graduation mark; this will reduce but not eliminate the variation in the amount of water added to each volumetric flask.

many steps in an analysis, and are estimated (see Equation 3) from the results of replicates: they can be minimised by good technique, but never eliminated.

The most important characteristic of an analytical result is usually the average or representative value for the level of the analyte in the matrix (a measure of location), and an estimate of the uncertainty involved in that representative value. A number of simple but important statistical concepts are involved in calculating and using such information.[30,31]

In most analyses, replicate measurements are carried out so that random errors can be estimated. In such cases the average value of the analyte level is normally expressed using the **arithmetic mean**, $\bar{x}$, (often simply called the **mean**) given by Equation 2:

$$\bar{x} = \frac{\sum_i x_i}{n} \tag{2}$$

where the x_i values are the n individual measurements, *i.e.* the sample size is n. An alternative measure is the **median**: if n is odd, the median is the middle value of the measurements when they are arranged in numerical order, while if n is even, the median is the average of the two middle measurement results, again when they are ordered. The median is widely used in non-parametric statistics (see Section 6.1).

Random errors are usually expressed as the sample standard deviation, s, given by Equation 3:

$$s = \left\{ \frac{\sum_i (x_i - \bar{x})^2}{n - 1} \right\}^{1/2} \tag{3}$$

Note that the term $(n - 1)$ is used in the denominator of this equation to ensure that s is an *unbiased estimate* of the population standard deviation, σ. (There is a general convention in statistics that English letters are used to describe the properties of samples, Greek letters to describe populations). The term $(n - 1)$ is the number of **degrees of freedom** of the estimate, s. This is because if $\bar{x}$ is known, it is only necessary to know the values of $(n - 1)$ of the individual measurements, as by definition $\sum_i (x_i - \bar{x}) = 0$. The square of the standard deviation, s^2, is known as the **variance**, and is a very important statistic when two or more sources of error are being considered, because of its additivity properties.

The standard deviation of the results is sometimes more meaningful if it is expressed in relation to the mean value, $\bar{x}$: it is usual to do this by calculating:

relative standard deviation (rsd) $= s/\bar{x}$;
or **coefficient of variation (%CV)** $= 100 \, s/\bar{x}$.

Note that the use of these two terms (rsd, CV) are not strictly adhered to and are interchanged in some literature.

In most analytical experiments where replicate measurements are made on the same matrix, it is assumed that the frequency distribution of the random error in the population follows the 'normal' or Gaussian form (these terms are also used interchangeably, though neither is entirely appropriate). In such cases it may be shown readily that if samples of size n are taken from the population, and their means calculated, these means also follow the normal **error distribution** ('the sampling distribution of the mean'), but with standard deviation $s/\sqrt{n}$ this is referred to as the **standard deviation of the mean** (sdm), or sometimes **standard error of the mean** (sem). It is obviously important to ensure that the sdm and the standard deviation s are carefully distinguished when expressing the results of an analysis.

Example: 6.4

The level of thyroxine in a single specimen of human blood was determined seven times with the following results: $x/\text{ng mL}^{-1}$

$$5.7, 7.0, 6.5, 6.2, 5.9, 6.9, 6.0$$

Here $n = 7$ (6 d.f.), $\bar{x} = 6.31$ ng mL^{-1}, $s = 0.50$ ng mL^{-1}, and CV $= 7.9\%$. The sdm $= 0.50/2.646 = 0.19$ ng mL^{-1}

It should be noted that when a single analyte is measured in different or separate matrices (*e.g.* the concentration of a protein in blood sera from different donors), the assumption of a Gaussian distribution might well not be valid. Other reasons for believing that this assumption is questionable are discussed below.

If the assumption of a Gaussian error distribution is considered valid, then an additional method of expressing random errors is available, based on **confidence levels**. The equation for this distribution can be manipulated to show that approximately 95% of all the data will lie within $\pm 2\,s$ of the mean, and 99.7% of the data will lie within $\pm 3\,s$ of the mean. Similarly, when the sampling distribution of the mean is considered, 95% of the sample means will lie within approximately $2s/\sqrt{n}$ of the population mean *etc.* (Figure 5).

In practice this implies that, if systematic errors are absent (an assumption always to be tested, never taken for granted!) we can be 95% confident that the true value of the analyte level lies within $\pm 2\,s/\sqrt{n}$ of the sample mean.

This gives us a **confidence interval**, for the true value, μ, whose extremities are the **confidence limits**.

$$\mu = \bar{x} \pm t(s/\sqrt{n}) \tag{4}$$

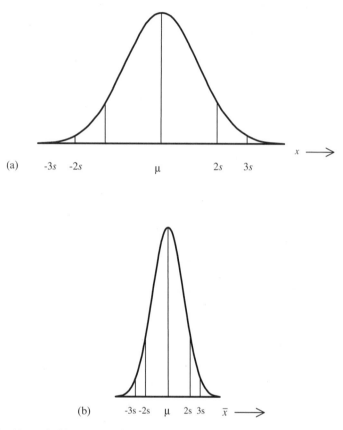

Figure 5 *Normal (Gaussian) distribution for (a) individual measurements, and (b)*
means of samples of measurements, with the areas enclosing 95% ($\pm 2\,s$) and
99.7% ($\pm 3\,s$) of the values shown in each case

The appropriate value of t, from the Student t distribution, depends on the
number of degrees of freedom ($n - 1$) and the level of confidence required.

6.3 Significance Testing

Coverage factor	**Robust statistical method**
Expanded uncertainty	**Significance tests**
Method of least squares	**Standard uncertainty**
Multivariate statistics	**True value**
Null hypothesis	**Type I error**
Power of a test	**Type II error**
Probability levels	**Uncertainty**
Regression of *y* on *x*	***y*-Residuals**

The use of **confidence intervals** and **limits** (see Section 6.2) is closely related to the concept of **probability levels**. When the true value for an analyte level is estimated using, *e.g.* 95% confidence limits, there is a 5% probability (usually written $P = 0.05$, *i.e.* using decimal rather than percentage notation) that, if all the underlying assumptions are valid, the true value will lie *outside* the confidence interval. This concept is widely used in statistical **significance tests**. Such tests are used to explore whether the mean of an experimental result differs significantly from a stated value (*i.e.* whether systematic errors might be present); whether the means and standard deviations of two (or more) samples differ significantly; whether the slope and intercept of a calibration graph differ significantly from 1 and 0 respectively; and so on. A typical significance test is carried out by:

(1) setting up a **null hypothesis**, given the symbol, H_0 which expresses the situation that the samples *etc.* being compared do not differ, *i.e.* could come from the same population;
(2) converting the experimental result to a test statistic, using standard formulae which depend on the test required;
(3) estimating with the aid of statistical tables or by direct calculation on a computer the probability that this test statistic would have been obtained if H_0 were valid;
(4) rejecting the **null hypothesis** if this probability is less than $P = 0.05$ (or any other value chosen by the analyst according to need) but retaining H_0 if the probability exceeds this value.

The chosen probability is the level at which an error may occur in the interpretation of the significance test outcome: if $P = 0.05$, the null hypothesis is rejected in 5% of the cases when in fact it should be retained. This is known as a **type I error**: the opposite or **type II error**, *i.e.* accepting a null hypothesis that should be rejected, can of course also occur, and it is normally possible to minimise both types of error simultaneously only by increasing n, the sample size.

The **power** of a significance test (see Equation 5) is a measure of the test's ability correctly to reject a false null hypothesis.

$$\text{Power} = 1 - \text{probability of a type II error} \qquad (5)$$

Some of the concepts used in defining confidence limits are extended to the estimation of **uncertainty**. The uncertainty of an analytical result is a range within which the **true value** of the analyte concentration is expected to lie, with a given degree of confidence, often 95%. This definition shows that an uncertainty estimate should include the contributions from all the identifiable sources in the measurement process, *i.e.* including systematic errors as well as the random errors that are described by confidence limits. In principle, uncertainty estimates can be obtained by a painstaking evaluation of each of the steps in an analysis and a summation, in accord with the principle of the additivity of variances (see above) of all the estimated error contributions: any systematic errors identified are expressed in a way that allows their contributions to be included in this addition process. If this ISO-recommended 'bottom-up' approach to uncertainty[32] is not feasible, then a 'top-down' approach, using the results of collaborative trials to get an overall estimate of the errors, may be used.[33] In either case, the uncertainty can be expressed in two ways. The **standard uncertainty**, u, expresses the result as if it were a standard deviation; the **expanded uncertainty**, U, is analogous to confidence limits and expresses the range within which the true value is expected to lie with a confidence of, *e.g.* 95%. These two definitions are related by the **coverage factor**, k, *i.e.* $U = u \times k$. As in the case of confidence limits, if U is a 95% expanded uncertainty value, then $k = 2$ (see Section 2.4).

As we have seen, results in many areas of instrumental analysis are obtained as a result of calibration experiments, in which the instrument signal given by a test material is compared by interpolation with the signals given by a range of standard materials. In many cases the calibration graph is a straight line, of the form $y = a + bx$, so it is necessary to calculate the values of a (intercept) and b (slope), bearing in mind that the instrument signals (y-values) given by the standards will be subject to random errors. Using the assumptions that the variations in these y-values are normally distributed and the same for all values of the exactly known standard concentrations (x-values), the slope and intercept are calculated using the **method of least squares**. This minimises the sum of the squares of the **y-residuals**, *i.e.* the y-direction distances between the points and the best line. This line is called the line of **regression of y on x**. This approach leads to the results shown in Equation 6 and Equation 7, where the individual standards give points (x_1, y_1) *etc.* and the centroid of the calibration points is $(\bar{x}, \bar{y})$:

$$b = \frac{\sum_i (x_i - \bar{x})(y_i - \bar{y})}{\sum_i (x_i - \bar{x})^2} \tag{6}$$

$$a = \bar{y} - b\bar{x} \tag{7}$$

If the graph defined by these a and b values, and with n calibration points, is then used to obtain the concentration, x_0 of a test material from m replicates of the instrument signal with mean y_0 the standard deviation, s_{x_0}, of this concentration is given by:

$$s_{x_0} = \frac{s_{y/x}}{b} \left\{ \frac{1}{m} + \frac{1}{n} + \frac{(y_0 - \bar{y})^2}{b^2 \sum_i (x_1 - \bar{x})^2} \right\}^{1/2} \tag{8}$$

where $s_{y/x}$, the standard error of the calibration points, is given by:

$$s_{y/x} = \left\{ \frac{\sum_i (y_i - \hat{y}_i)^2}{n - 2} \right\}^{1/2} \tag{9}$$

In Equation 9 the $\hat{y}_i$ values are the 'fitted' y-values, *i.e.* the values of y on the best line at the x-value of the standard. More complex equations are necessary if weighted regression has to be used, *i.e.* when the y-direction error is not the same at all values of x. Moreover, many analytical methods give curved calibration plots, fitted for example by polynomial graphs:

$$y = a + bx + cx^2 + dx^3 + \ldots \tag{10}$$

These topics are dealt with in detail in many texts.[34,35]

Numerous studies over recent years have shown that, in analytical and other experimental sciences, the assumption of a purely Gaussian error distribution for replicate measurements may not be wholly appropriate. A more typical distribution may be one that is superficially similar to the Gaussian, but is 'heavy-tailed', *i.e.* it is symmetrical about the mean but has rather more measurements than expected which are distant from the mean value. Such an outcome might result from the presence of gross errors, *i.e.* the heavy tails represent potential **outliers** (see Section 6.1). Alternatively the measurements might be performed in practice by different individuals, on different equipment or under different laboratory conditions. In such cases the observed error distribution might in reality be the sum of several separate Gaussian distributions, with similar means if systematic errors are absent, but different standard deviations. These problems are now frequently addressed by the use of **robust statistical methods**,[36] which have the property of down-weighting by various approaches (rather than ignoring entirely) measurements which are distant from the mean. Robust regression methods are also popular.[37] All these techniques circumvent the problems of potential outliers, but in contrast to non-parametric methods do not discard entirely the importance of the underlying Gaussian pattern of errors. Many robust methods require iterative calculations, but these are readily performed on personal computers, and some of the major commercially available suites of statistical software are now offering such techniques.

Finally it is important to note that modern analytical equipment frequently offers opportunities for measuring several or many characteristics of a material more or less simultaneously. This has encouraged the development of **multivariate statistics** methods, which in principle permit the simultaneous analysis of several components of the material. Partial least squares methods and principal component regression are examples of such techniques that are now finding extensive uses in several areas of analytical science.[38]

7 Acronyms

ANSI	American National Standards Institute
AQL	Acceptable Quality Level
ASTM	American Society for Testing Materials
BCR	Bureau Communautaire de Référence
BIPM	Bureau International des Poids et Mesures
BS	British Standards
BSI	British Standards Institution
CEN	European Committee for Standardization
CRM	Certified Reference Material
EN	European Standards
GLP	Good Laboratory Practice
GMP	Good Manufacturing Practice
ISO	International Organization for Standardization
IUPAC	International Union of Pure and Applied Chemistry
LAL	Lower Action Limit
LIMS	Laboratory Information Management System
LOD	Limit of Determination
LOQ	Limit of Quantification
LWL	Lower Warning Limit
NIST	National Institute of Standards and Technology
OIML	International Organization of Legal Metrology
PT	Proficiency Testing
QA	Quality Assurance
QC	Quality Control
RM	Reference Material
SI	International System of Units
SOP	Standard Operating Procedure
SRM	Standard Reference Material
TQM	Total Quality Management
UAL	Upper Action Limit
UKAS	United Kingdom Accreditation Service
UWL	Upper Warning Limit
VIM	International vocabulary of basic and general terms in metrology

Statistical Symbols Used in Quality Assurance

$\bar{x}$	The arithmetic mean value of a sample set of data points
μ	The mean value of an entire population of data points
s	The sample standard deviation of a set of data points
σ	The standard deviation of an entire population of results
s^2	Variance
rsd	Relative standard deviation
CV	Coefficient of Variation
r	Repeatability
R	Reproducibility
v	The numbers of degrees of freedom
t	The tabulated Student t value for a given level of confidence with v degrees of freedom
$u(x_i)$	Standard uncertainty, uncertainty of the result x_i of a measurement expressed as a standard deviation
$u_c(y)$	Combined standard uncertainty of y
k	Coverage factor
U	Expanded uncertainty

8 References

1. L. H. Keith, *Environmental Sampling and Analysis, a Practical Guide*, Lewis Publishers, 1991.
2. N. T. Crosby and I. Patel, *General Principles of Good Sampling Practice*, Royal Society of Chemistry, 1995, ISBN 0 85404 412 4.
3. 'Sampling' in *Encyclopaedia of Analytical Science*, Ed. A. Townshend, Academic Press, 1995.
4. W. Horwitz, 'AOAC no longer analyzes samples – we analyze test samples', *Inside Lab. Management*, 2000, **4**, 3, ISSN 1092-2059.
5. International Standards Organization, *International Vocabulary of Basic and General Terms in Metrology*, Blackwell Science, 1993, ISBN 9 26701 705 1.
6. The definition for standard operating procedure from the US Food and Drug Administration web site – www.fda.gov
7. Coordinating Editors M. Sargent and G. MacKay, *Guidelines for Achieving Quality in Trace Analysis*, Royal Society of Chemistry, 1995, ISBN 0 85404 402 7.
8. A. D. McNaught and A. Wilkinson, *Compendium of Chemical Terminology – IUPAC Recommendations*, Blackwell Science, 1997, ISBN 0 86542 684 8.
9. *Codex alimentarius commission*, ALINORM 97/23A, Appendix 3.
10. T. Farrant, *Practical Statistics for the Analytical Scientist*, Royal Society of Chemistry, 1997, ISBN 0 85404 442 6.
11. R. Grant and C. Grant, *Grant and Hackh's Chemical Dictionary*, 5th Edition, McGraw-Hill Inc, 1987, ISBN 0 07024 067 1.
12. EURACHEM Guide, *The Fitness for Purpose of Analytical Methods*, LGC, 1998, ISBN 0 94892 612 0.
13. 'Nomenclature in evaluation of analytical methods, including detection and quantification capabilities' (IUPAC Recommendations 1995), *Pure Appl. Chem.*, 1995, 1699–1723.
14. Draft International Harmonisation of Pharmacopoeias, 'Text on validation of analytical procedure', *Pharmeuropa 5*, 1993, No 4, 341.
15. E. Prichard with G. MacKay and J. Points, *Trace Analysis: a structured approach to obtaining reliable results*, Royal Society of Chemistry, 1996, ISBN 0 85404 417 5.
16. ISO Guide 30: 1992, Terms and definitions used in conjunction with reference materials.
17. D. J. Bucknell, *Referencing the Reference Materials – the Ultimate Act of Faith, VAM Bulletin*, Issue No: 12, Spring 1995, ISSN 0957-1914.
18. CITAC Guide 1: *International Guide to Quality in Analytical Chemistry, An Aid to Accreditation*, LGC, 1995, ISBN 0 94892 609 0.

19. F. M. Garfield, *Quality Assurance Principles for Analytical Laboratories*, 2nd Edition, AOAC International, 1991, ISBN 0 93558 446 3.
20. G. Kateman and L. Buydens, *Quality Control in Analytical Chemistry*, 2nd Edition, John Wiley & Sons, 1993, ISBN 0 47155 777 3.
21. E. Prichard, *Quality in the Analytical Chemistry Laboratory*, John Wiley & Sons, 1995, ISBN 0 47195 541 8.
22. BS EN 45020: 1993, *Glossary of terms for Standardization and related activities.*
23. Information from UKAS – www.quality-register.co.uk
24. J. K. Taylor, *Quality Assurance of Chemical Measurements*, Lewis Publishers, 1987, ISBN 0 87371 097 5.
25. ISO 2859-1 ($\equiv$ BS 6001: Part 1), *Specification for sampling plans indexed by acceptable quality level (AQL) for lot by lot inspection*, 1989.
26. R. E. Lawn, M. Thompson and R. F. Walker, *Proficiency Testing in Analytical Chemistry*, Royal Society of Chemistry, 1997, ISBN 0 85404 432 9.
27. ISO Guide 33: *Uses of Certified Reference Materials*, 1989.
28. M. Thompson and R. Wood, *J. AOAC Int.*, 1993, 926.
29. J. N. Miller, *Analyst (Cambridge)*, 1993, **118**, 455.
30. J. N. Miller and J. C. Miller, *Statistics and Chemometrics for Analytical Chemistry*, 4th Edition, Prentice Hall, 2000, ISBN 0 13022 888 5.
31. W. P. Gardiner, *Statistical Analysis Methods for Chemists*, Royal Society of Chemistry, 1997, ISBN 0 85404 549 X.
32. *Guide to the Expression of Uncertainty in Measurement*, ISO, 1993, ISBN 9 26710 188 9.
33. *Quantifying Uncertainty in Analytical Measurement*, 2nd Edition; see Eurachem web site for how to obtain a copy of the guide – www.vtt.fi/ket/eurachem
34. J. N. Miller, *Analyst (Cambridge)*, 1991, **116**, 3.
35. N. R. Draper and H. Smith, *Applied Regression Analysis*, 3rd Edition, John Wiley & Sons, 1998, ISBN 0 47117 082 8.
36. Analytical Methods Committee, *Analyst (London)*, 1989, **114**, 1693.
37. P. J. Rousseeuw and A. M. Leroy, *Robust Regression and Outlier Detection*, John Wiley & Sons, 1987, ISBN 0 47185 233 3.
38. D. L. Massart *et al.*, *Handbook of Chemometrics and Qualimetrics*, Parts A and B, Elsevier Science, 1997, ISBN 0 44482 854 0.

Subject Index

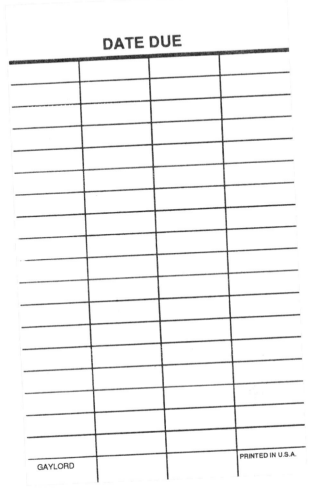